UNDERSTANDING DIGITAL ETHICS

Rapid changes in technology and the growing use of electronic media signal a need for understanding both clear and subtle ethical and social implications of the digital, and of specific digital technologies. *Understanding Digital Ethics: Cases and Contexts* is the first book to offer a philosophically grounded examination of digital ethics and its moral implications. Divided into three clear parts, the authors discuss and explain the following key topics:

- Becoming literate in digital ethics
- Moral viewpoints in digital contexts
- Motivating action in digital ethics
- Speed and scope of digital information
- Moral algorithms and ethical machines
- The digital and the human
- Digital relations and empathy machines
- Agents, autonomy, and action
- Digital and ethical activism.

The book includes cases and examples that explore the ethical implications of digital hardware and software including videogames, social media platforms, autonomous vehicles, robots, voice-enabled personal assistants, smartphones, artificially intelligent chatbots, military drones, and more.

Understanding Digital Ethics is essential reading for students and scholars of philosophical ethics, those working on topics related to digital technology and digital/moral literacy, and practitioners in related fields.

Jonathan Beever is Assistant Professor of Ethics and Digital Culture with the Department of Philosophy and the Ph.D. Program in Texts and Technology at the

University of Central Florida, USA. He is co-editor (with N. Morar) of *Bioethics, Science, and Public Policy* and (with V. Cisney) of *The Way of Nature and the Way of Grace: Philosophical Footholds on Terrence Malick's The Tree of Life*.

Rudy McDaniel is Professor of Games and Interactive Media in the Nicholson School of Communication and Media and Director of the School of Visual Arts and Design at the University of Central Florida, USA. He is co-author (with J.D. Applen) of *The Rhetorical Nature of XML* and (with Joseph Fanfarelli) *Designing Effective Digital Badges: Applications for Learning*, both published by Routledge.

Nancy A. Stanlick is Professor of Philosophy and Associate Dean in the College of Arts and Humanities at the University of Central Florida, USA. She has published five books: *Philosophy in America: Volumes I and II* (co-authored and co-edited with Bruce Silver), *Asking Good Questions: Case Studies in Ethics and Critical Thinking* (with Michael Strawser), *The Essential Leviathan: A Modernized Edition*, and *American Philosophy: The Basics* (also available from Routledge).

UNDERSTANDING DIGITAL ETHICS

Cases and Contexts

Jonathan Beever, Rudy McDaniel and Nancy A. Stanlick

LONDON AND NEW YORK

First published 2020
by Routledge
2 Park Square, Milton Park, Abingdon, Oxon OX14 4RN

and by Routledge
52 Vanderbilt Avenue, New York, NY 10017

Routledge is an imprint of the Taylor & Francis Group, an informa business

© 2020 Jonathan Beever, Rudy McDaniel and Nancy A. Stanlick

The right of Jonathan Beever, Rudy McDaniel and Nancy A. Stanlick
to be identified as authors of this work has been asserted by them in
accordance with sections 77 and 78 of the Copyright, Designs and
Patents Act 1988.

British Library Cataloguing-in-Publication Data
A catalogue record for this book is available from the British Library

Library of Congress Cataloging-in-Publication Data
A catalog record has been requested for this book

ISBN: 978-1-138-23333-1 (hbk)
ISBN: 978-1-138-23334-8 (pbk)
ISBN: 978-1-315-28213-8 (ebk)

Typeset in Bembo
by Deanta Global Publishing Services, Chennai, India

CONTENTS

FIGURES

ACKNOWLEDGMENTS

Jonathan thanks students in his "Digital Ethics" graduate seminars in Spring 2016 and Fall 2018 and his co-authors on this project: thoughtful engagement with topics and concepts relevant to this book helped to strengthen the arguments and examples we develop. He offers special thanks to his family (two- and four-legged) for their amazing support and encouragement in this and all projects.

Rudy thanks his colleagues at UCF for ideas, inspiration, and discussion around many of the ethical issues and examples discussed in this book. Nancy and Jonathan were tremendous partners on this project and he learned much from both of them. He appreciates the time provided by Dean Jeffrey Moore to work on this book. Lastly and most importantly, he thanks his family: Carole, Brighton, Ben, and Becca, for their constant support and encouragement.

Nancy thanks Jonathan and Rudy for their expertise and ideas. She also appreciates the research time provided to work on this book by Dean Jeffrey Moore of the College of Arts and Humanities at the University of Central Florida. And as always, I thank my family for their patience while I sequestered myself in a little room for untold hours in front of computer screens. To Teddy, the cat: Thanks for walking on the keyboard to keep me amused with error correction.

AN INTRODUCTION

Digital ethics is a way of thinking about the ethical implications of "the digital," or digital technologies and the cultures they co-constitute. In some ways it is akin to other emergent fields of applied ethics, like nano-ethics, which spawned its own journal and asked self-critical questions about whether emergent bio-technology brought along with it its own novel ethical problems. Emergences like these are circadian, periodically drawn to the surface by a complex array of factors to make noise, connect with other ideas, and—potentially—spawn a new generation of questions and concepts.

Digital ethics is at a critical space of emergence thanks to the rapid rise of digital technologies and the slower growth of social concerns about them. Digital technologies are nearly everywhere in contemporary life across the globe. Their effects are felt even by those who have little or no access to them for their personal use. Digital technologies and media, because of their usefulness and pervasiveness, affect our lives, social structures, policies, governments, and families. These effects have significant ethical implications that are easy to overlook, especially given the pervasiveness and embeddedness of the digital in our everyday lives.

Given the rapid changes in technology and the heavy use of electronic media in several diverse areas (e.g., entertainment, business, training, education) there is a need for an ever-clearer understanding of the ethical and social implications of the digital—concerns that show up across contexts. Politicians find themselves in the embarrassing position of being caught using social media to send sexual imagery to lovers, and social media in general is proving a scandalous affair in politics (Shear, 2011). Other public officials are using social media in ethically questionable ways to gather information or to influence decision-makers (Stutzman, 2015). The military is investigating drone technologies and "lethal autonomous weapons systems" (*Nature*, 2015); some are calling for better understandings of the implications of robot-computer warfare and the "new ethics of war" (Gordon, 2015). People are under a constant barrage of imagery and noise from electronic sources that puts at risk members of younger generations who have been dubbed "babies with superpowers" (James et al., 2009, p. 18) for their combination of naiveté and digital reach.

Given this vast scope of impact, a clearer understanding of the nature and scope of digital ethics is certainly valuable, although not easy to achieve. Doing digital ethics requires engaging ethical decision-making strategies as well as understanding the nature of the technologies and processes that constitute the digital itself: moral literacy alongside digital literacy. Indeed, nearly everyone believes that he or she has a basic and working grasp of ethics already. We live our lives in both ordinary and more complicated moments making moral decisions that have an impact on ourselves, others, and sometimes even on society. We may act based on adages such as "honesty is the best policy," "respect your elders," and "do the right thing." However, very few people have the time or the inclination to look more deeply at the decisions they make and the impact of the resulting actions. Even a critical comment on Facebook or a tweet sent without much thought can have profound impacts on others, as evidenced by high profile news stories such as the Justine Sacco incident, in which a corporate communications director was taken to task by Internet communities, ultimately losing her job, thanks to her insensitive and offensive comments that she thought were appropriate for the medium: "Going to Africa. Hope I don't get AIDS. Just kidding. I'm white!" (Ronson, 2015). Even if one acts within more generally accepted conventions of decency (e.g., "if you don't have anything nice to say, don't say anything at all"), it is rare that ethical reasoning, decision making, and acting are so simple as to require nothing but comforting—and unfortunately often inapplicable—adages or general rules. Similarly, nearly everyone believes that they have a clear idea of what the digital is, or at least what digital technologies are. But it is rare that our intuitive understanding of what the digital is—in terms of its nature, scope, and impact on our lived experiences—is coherent. The argument we make in this book is that only by coupling knowledge of the digital with analysis of ethical issues around it can we really get at a "digital ethics."

For example, consider the processes of information gathering, reasoning, and decision making that are involved in something as seemingly banal as whether one ought to purchase a tablet computer or other Internet-connected device for her 12-year-old son. Tablets, laptop computers, and Android and Apple phones are ubiquitous in industrialized societies. Because of this popularity, a mother may simply succumb to the persistent requests of her son to have an Internet-connected device through which he can communicate with his friends via email, text messaging, Facebook, or other social media like Twitter or Instagram. On this device, the son explains to mom, he will be able to look up information for homework assignments for school. On the surface, the mother's final decision may sound simple and innocuous: after all, millions of people own and use such devices, many of whom are young children, and the devices can be useful and beneficial in numerous ways. For example, a parent will be able to contact a child at school or when away from home; the child will be able to do homework more quickly and efficiently; and even entertainment available online will be a benefit to the child, allowing him to watch TV shows, movies, and educational content on demand. These are certainly benefits of the use of this type of digital technology, but the decision, while seemingly simple, is not always so.

The potential for good that may come from the use of digital technology combines with the potential for problems having negative consequences on its owner and others. While her son may use a new Internet-connected tablet to complete homework assignments, he might also find himself drawn into websites or social media interactions of questionable value and of moral concern. Many websites contain advertisements for everything from sugary snacks to credit monitoring services and from mundane diaper-delivery to much more risqué "adult" services. Should the parent control and limit the use of the device by using parental control on the device, limiting access to the Internet, or in some other way protecting her son from such available uses of digital technology? Sometimes, and perhaps often, children are savvier about the use of digital technology than their parents. A child might override parental control or use the Internet at public Wi-Fi "hotspots," effectively negating the parent's attempts to minimize access and exposure to questionable or dangerous content. Perhaps, then, the decision whether to purchase a tablet for one's son extends beyond the benefits to be gained from its possession and use and moves further into questions regarding what kind of moral education the parent ought to provide to her children. From this example, it is clear enough that moral issues are not all completely black and white, and that not every moral decision is so simple to solve as thinking that a good consequence, or a bad one, is the only result that may occur. The decision to buy a 12-year-old a tablet computer might be made based on the good to which it can be put in advancing her son's education without consideration or without knowledge of the questionable or bad results that might also attend its purchase.

In part because our modern digital world facilitates relationships through technology, we encounter ethical issues like this in our everyday digital lives in virtually every aspect of our actions, beliefs, behaviors, preferences, and conversations. Deciding what to do, observing who will be affected, considering the possible or likely results of our actions, and even understanding the way in which we have come to a decision are all actions with moral dimensions. It is no different whether a person is trying to decide on the best or right course of action in donating money to a charitable cause (perhaps using a digital tool like Charity Navigator) or making a decision in an online business context (perhaps by choosing a worthy startup to endorse using Kickstarter) that has the potential to affect the lives and livelihoods of employees. Digital technology, like every other aspect of our lives, presents us with dilemmas that have far-reaching implications for other people, nonhuman animals, the environment, government, the conduct of business, and our individual and social existences.

Our goal with this book is to prepare thoughtful people like you for action to engage situations and structures you identify as unethical and to engage in productive work on questions of digital ethics. This model of identifying ethical issues, analyzing them through richly pluralistic theory, and motivating our readers to act in the world toward the ends to which they have reasoned, frames our approach to digital ethical literacy.

Our Approach

We start in Chapter 1 by analyzing the intersections of digital literacy and ethical literacy that are required to understand digital ethics. Digital literacy is the process of coming to identify, reason through, and act in relation to digital technologies, media, and environments. It is a process of developing specialized knowledge and skills through practice and engagement, much like the ways one becomes literate in traditional terms of reading and writing. Similarly, becoming morally literate requires identification and understanding of moral viewpoints and ethical concepts, an understanding of how to reason well about and with them, and motivation to act on those reasoned decisions. We review major ethical theories and advocate for a pluralistic approach in Chapter 2, then turn our focus to motivation in Chapter 3. In both chapters, we argue that taking digital ethics to be this process of becoming literate requires both epistemic or knowledge-based inquiry as well as ethical or values-based inquiry.

We then dedicate a chapter each to key properties of the digital, including *distributedness*, *procedurality*, and *embeddedness*, before turning to their implications for agency and empathy. The reproducibility and transferability of the digital, in its various forms, stems from the *distributed* nature of digital information that enables unlimited one-to-one reproductions. This is the topic of Chapter 4. Such properties have had profound implications, ranging from ethical questions about replication to deep ontological questions about authenticity, lack of scarcity, and modes of representation. The binary nature of digital information structures not only the *programmability* of digital media but also its *procedurality*, a second set of key properties of the digital that challenges traditional ethical concerns grounded in ontological conceptions of the uniqueness of media objects. We discuss procedurality, along with programmability and the possibility of moral algorithms, in Chapter 5. Finally, given that individuals today are often networked to one another extensively through social media and informational technologies such as Facebook, Twitter, Instagram, and Reddit, it is not surprising that embeddedness as a property of the digital has propagated value systems that are shared by many users. Ubiquitous computing (the ways digital content is constantly at hand), identity and anonymity, and organizational culture are all issues related to the *embeddedness* of the digital. We use contemporary examples dealing with embeddedness to frame the concepts and problems we review in Chapter 6.

Our attention turns next to two implications of these properties that we see as being of fundamental ethical importance. First, the flows of digital information embedded in social and personal contexts have the potential to enhance the ways we connect to others; that is, to enhance our capacity for "empathy." However, those same flows of digital information face ethically negative challenges as well: they can lead to *desensitization* to events and states of affairs that would normally trigger an empathetic response. With so much unfiltered information available, becoming desensitized to the suffering of others is a central ethical concern relevant to digital ethics. We discuss these issues thoroughly in Chapter 7. Second, digital technologies, media, and modes of interaction *re-situate agency* by framing users of digital media as designers, producers, and consumers *of it* rather than mere causal actors *upon it*. In a world shaped by digital information flows, agency is transformed along several dimensions and we discuss changes to agency regarding sociality and time in Chapter 8. Imagine the online gamer whose digital avatar affects whole environments wholly distinct from the gamer herself. Imagine the drone operator, whose digital agency is reflected thousands of miles away by the engagement of the enemy in warfare. In the analog world, such ethical issues were much more difficult to conceive: agency and the ethicality of action were linked tightly to physical causal interaction.

In the final chapter of our book, Chapter 9, we think through the ways that understanding ethical implications can and should have practical implications for policy and practice. We situate a fundamental problem of digital ethics in practice—the move from developing reasons to motivating action—and suggest that digital technologies offer novel and robust ways of acting with effect, which has both positive and negative ethical implications. We examine existing policies and practices for action related to issues of digital ethics as a landscape of existing resources and spaces for future work.

In the conclusion, we tie together the various case-based analyses of themes and concepts in digital ethics through an exploratory discussion of what the future of digital ethics might hold and what skills are required to sustain understanding and action within it. We return to the relationship between digital and moral literacy in support of our overarching argument that ethics, and in particular digital ethics, requires communities of engagement.

Overall, our approach relies on the balance between conceptual and theoretical work and its situation within contemporary examples that are opportunities for ethical engagement and analysis. Carefully designed case studies provide effective modes by which to engage in conversation and motivate students to act on ethical decisions by cultivating habits of inquiry, and developing moral hope, purpose, and courage by which to act. Yet, developing case studies that do this work of balancing theory is not easy, so we have included a practically oriented pedagogical appendix as a final reference. There, we discuss the process of developing and conducting case studies in digital ethics as a means of motivating students to engage in digital ethical literacy. The guidelines we develop form a template that readers might wish to develop in their own thinking about digital ethics.

In sum, *Understanding Digital Ethics* examines the complex nature of digital technologies, media, and modes of communication and the many ethical issues that arise from it. Through concepts and case studies we develop, our book combines knowledge-building exercises about "the digital" with both traditional and contemporary ethics and social philosophical examples and content. Digital technologies, media, and modes enable richer communication and greater access to information. But, at the same time, they do so by changing the context and content of our interactions. Developing digital literacy alongside ethical literacy is required to understand digital ethics. This book will enrich the dialogue about digital ethics by drawing out an explicit social justice angle that focuses attention on the ways in which the digital enables and restricts access and interaction. The case studies included draw out relevant conditions and values through the moral literacy framework. Our goal is to offer a set of diverse and rich examples, both real and fictional, to stand as an instructional framework by which readers can build their own case studies based on their unique digital experiences.

References

Gordon, N. (2015, January 23). Drones and the new ethics of war.*Common Dreams.* Retrieved from http://www.commondreams.org/views/2015/01/23/drones-and -new-ethics-war.

James, C., Davis, K., Flores, A., Francis, J.M., Pettingill, L., Rundle, M., & Gardner, H. (2009). *Young people, ethics, and the new digital media: A synthesis from the Goodplay Project.* Cambridge, MA: MIT Press.

Nature (2015, May 27). Robotics: Ethics of artificial intelligence. Retrieved from http://www.nature.com/news/robotics-ethics-of-artificial-intelligence-1.17611.

Ronson, J. (2015, February 12). How one stupid tweet blew up Justine Sacco's Life. *The New York Times.* Retrieved from http://www.nytimes.com/2015/02/15/magazine/ how-one-stupid-tweet-ruined-justine-saccos-life.html.

Shear, M.D. (2011, June 2). For politicians, social media holds promise and peril. *The New York Times.* Retrieved from http://thecaucus.blogs.nytimes.com/2011/06/02/ for-politicians-social-media-holds-promise-and-peril.

Stutzman, R. (2015, October 6). Florida Supreme Court reprimands judge Debra Krause for violations. *Orlando Sentinel.* Retrieved from http://www.orlandosentinel.com/ news/os-judge-debra-krause-reprimand-20151006-story.html.

PART I
Ethical and Digital Literacy

1

BECOMING LITERATE IN DIGITAL ETHICS

Literacy in digital ethics is often sadly lacking in both contemporary discourse and the design of technologies that support our everyday conversations and interactions in digital spaces. Consider just one of many recent examples from which we could pull to illustrate this claim. Facebook, and its CEO Mark Zuckerberg, continue to make national headlines for ethical issues ranging from data privacy, to propagation of fake news, to enabling Russian interference in the 2016 presidential election. A reporter pointed clearly to the underlying problem: "The fact is that Facebook's underlying business model itself is troublesome: offer free services, collect user's private information, then monetize that information by selling it to advertisers or other entities" (Francis, 2017). Like so many other big data corporate entities, Facebook's ethical issues are rooted in its business model. But the ethical problems Facebook faces were made possible by failures of its creators. Mark Zuckerberg and Facebook lead to a fascinating case about the importance of ethical and digital literacy.

One could argue that Zuckerberg has largely failed to be sensitive to or to identify the ethical implications of his corporate creation, which he has previously argued is a neutral technology platform, not a media company responsible for its content (Constine, 2016). Of course, he has a deep and meaningful understanding of the technical platform. Yet, until recently perhaps, he has neglected to consider the broader ethical implications of actions that platform makes possible. Only in March 2018 did Zuckerberg publicly start to identify even basic ethical considerations (see Swisher & Wagner, 2018). And Facebook's slow turn, starting in 2016 after Russian interference in the U.S. presidential election, to tackling the problem of fake news is another indicator of its recognition of its ethical responsibility (e.g., Boorstin, 2016; Schroeder, 2019). This turn marks Facebook's joining of the ongoing conversation between the public, digital information industries like Facebook, and other thought leaders on how to reason carefully about these ethical implications. And all this reasoned thinking is targeted toward whether and what kind of policy and practice changes must be made in order to resolve ethical issues. Should information companies like Facebook be federally regulated? Should users be better informed of how their information is being used? We will engage further with Facebook as a digital ethics case later in this book. But to us, Facebook's controversy reflects the importance of the processes of digital and moral literacy in understanding digital ethics—the topic of this chapter.

Think about how you became literate. Think next about the implications of *failing* to become literate. Our guess is that your stories are much like our own. We became literate through a developmental process, learning first basic skills and scaffolding those up through experience and habit, guided by mentors, teachers, friends, and family who shared their expertise and experiences with us. And when we think about *failures* to become literate, we think of problems of access, inequality, and injustice. And just like one becomes literate in the context of reading and writing, we become literate in the context of ethics and the digital, too.

In this chapter we develop the claim that digital ethics requires engaging the intersection of *moral literacy* and *digital literacy*. Important problems lie at this intersection, including identifying novel ethical issues regarding emerging technologies, analyzing problems about the nature of stakeholders and their autonomy, and understanding a process of ethical decision making about digital issues. But let us first start with some philosophy.

The (Self-Driving) Trolley Problem

One of the most famous ethical thought experiments in moral philosophy is known as the "trolley problem," dealing with questions of technological control, agency, and moral responsibility. The problem originated in the work of philosopher Philippa Foot in 1967 (Marshall, 2018) in the context of abortion debate, and has since been adapted widely for numerous different applications (e.g., Thomson, 1985). In one version of this hypothetical scenario, there is a runaway trolley car and an individual with access to a track-switching lever. For some reason, there are five innocent persons tied to the main track and one other innocent person tied to a secondary track. The train can be diverted to the secondary track if someone pulls a lever in the train yard. If the individual witnessing this imminent disaster does nothing, the trolley will kill five people tied to the main track. If they pull the lever, the trolley is diverted to a secondary track where it kills one person. The moral implications of this thought experiment, and its variations, have been debated by philosophers for decades. Around it revolve questions of vital importance: How far does our agency extend? What are the limits of our moral responsibility? To whom (or what) do we owe moral concern?

In the context of the digital, surrounded as we are by digital information and digital technologies that mediate endless flows of information, this thought experiment takes on renewed life. Indeed, coordinating the movement of human beings safely and efficiently has been a fundamental problem of human society for hundreds of years, so it is no surprise that many of our analog and digital technologies are directly or indirectly related to transportation systems. In fact, some of our earliest digital technologies, including the electric telegraph, were adopted more quickly and disseminated more widely due to their ability to solve transportation challenges. The early railroad systems in the U.S. and Britain, for instance, presented novel complications that required new technologies to address. As Gere (2008) explains, "The electric telegraph and Morse code were adopted as a solution for the 'crisis of control' in what was then possibly the most complex system ever built, the railways. Both in Britain and the U.S. the early railways were troubled by large numbers of accidents as well as problems with efficiency, mostly owing to the difficulty of coordinating different trains on the same line" (p. 35). With the telegraph's ability to rapidly (at the time) send information from one part of the track to another, many miles away, technology was able to address a fundamental problem of complexity tied to this emerging transportation system.

Dealing with this "crisis of control" described by Gere led to technological advancements in our railway systems, which was a largely positive outcome (although an analysis of labor practices in railway construction introduces several other moral problems outside the scope of this book). Too much control can also lead to moral anxiety, though, as revealed by the scenarios presented in the trolley problem. In modern times, a third scenario has emerged in that we are more often faced with the crisis of *not being in control*. For instance, our transportation systems are becoming more self-reliant, with technologies like autopilot, GPS navigation, and real-time safety mechanisms increasingly moving control from human operators to complex hardware and software systems. Automation is perhaps most frequently associated with the autopilot feature used in modern commercial airliners, but we are now seeing viable and operational automation within the commuter and passenger vehicle industry. These types of vehicles are often referenced in the media as "self-driving" or "autonomous" vehicles.

Autonomous vehicles can function without continuous human input. Simply put, they are self-driving cars and trucks. We have long seen examples of autonomous vehicles in fiction—the iconic self-driving cars in films such as *Minority Report* (Spielberg et al., 2002) provided visually compelling examples of these technologies long before we observed the clunky prototypes that now appear in the real world doing tasks like mapping streets and roads for GPS applications. Today, autonomous vehicles are big business, with one Boston consulting group finding over $80 billion invested in autonomous vehicle technology since 2014 and projecting a 60 percent potential saving in fuel cost for consumers who may one day use shared autonomous vehicles (Worland, 2017). Since we know that fuel is a finite resource, there is much attention being paid to autonomous vehicles as new technologies that will be "better" versions of the vehicles we use today. They can be better by being safer, using less fuel, making fewer mistakes than human drivers, and requiring less infrastructure, like parking spots and parking garages. Imagine how these vehicles might drop their owners off at work, go home to the garage to recharge, and then circle back around at the end of the day to pick them back up, for example.

In order to function, these vehicles use algorithms that learn and adapt to continuously changing parameters on the road and in traffic patterns. These functions and features allow the vehicles to behave autonomously, an advanced state of operation dependent on the ability for technology to be automated. Manovich (2013) notes that automation is one of the fundamental properties of computing. In Manovich's words, "As long as a process can be defined as a finite set of simple steps (i.e. as an algorithm), a computer can be programmed to execute these steps without human input" (p. 128). While such automation seems innocuous in many applications, such as controlling the temperature in an electronic toaster or setting the time for a recurring alarm in a digital alarm clock, the ethical implications become more significant when considering certain digital technologies, such as the driverless vehicles we are discussing here. For example, consider the reduced autonomy of the people who are riding in such vehicles. The individuals lack control, the vehicle may have a limited range of travel, and there is something, according to some people, that is simply "wrong" with autonomous vehicles. They are seen to impede autonomy and take part of the driver's and the passengers' freedom away. When you are in control of where the vehicle goes, there is a sense of responsibility that comes with this control. When a computer does this work for you, both autonomy and responsibility are significantly reduced.

When automation is combined with artificial intelligence, decisions that were previously made by human beings are offloaded to computer software. The combination of automation and AI-based decision making is particularly troublesome to some, as evidenced by articles such as Hill's (2016) essay about self-driving cars. In this piece, Hill notes that self-driving cars are a reality as of the year 2016 and routinely send prototypes throughout the streets of Silicon Valley (albeit with a human backup operator for emergency purposes). However, ponders Hill, what happens when the autonomous vehicle is faced with a "no-win" scenario in which a collision is imminent and the vehicle must decide which lives take priority in the upcoming accident? This is similar to the trolley problem discussed above. Hill's article is drawn from an article published in *Science* in which the authors pose the dilemma even more directly: "Autonomous Vehicles (AVs) should reduce traffic accidents, but they will sometimes have to choose between two evils—for example, running over pedestrians or sacrificing itself and its passenger to save them" (Bonnefon, Shariff, & Rahwan, 2016, p. 1573). Indeed, when surveyed about such technologies, 1,928 survey participants admitted that while they agreed with a utilitarian decision-making approach in autonomous vehicles in which overall casualties are minimized even at the expense of the vehicle's passengers, they would prefer to purchase vehicles that would "protect their lives by any means necessary" (Hill, 2016, para. 4).

Even more fascinating are the cultural variations in the responses to updated versions of the trolley problem. In one such update, the MIT Media Lab created a "Moral Machine" (MIT Media Lab, n.d.) where users were allowed to "switch" the programming of an autonomous vehicle. Users were asked to "decide whether to, say, kill an old woman walker or an old man, or five dogs, or five slightly tubby male pedestrians" (Marshall, 2018, para. 2). The Moral Machine collected responses from 39.6 million decisions in 10 different languages from millions of people in 233 different countries and territories (Marshall, 2018, para. 1). The results diverged and depended on cultural norms and values. As Marshall explained (2018, para. 3):

> participants from eastern countries like Japan, Taiwan, Saudi Arabia and Indonesia were more likely to be in favor of sparing the lawful, or those walking with a green light. Participants in western countries like the US, Canada, Norway, and Germany tended to prefer inaction, letting the car continue on its path. And participants in Latin American countries, like Nicaragua and Mexico, were more into the idea of sparing the fit, the young, and individuals of higher status.

This example reveals that engaging digital ethics is a process not only inexorably linked to our own values and bound up in our own interpretations of the world, but one that also draws deeply from our cultural backgrounds, expectations, and ideologies. A 2018 *Nature* essay included the crux of the issue in its headline: "Moral choices are not universal" (Maxmen, 2018). Although it seems like ethically training vehicles to understand this and make better decisions in these difficult circumstances would be at the forefront of these companies' minds, sadly that is not the case because our technologies are not yet sophisticated enough. Marshall (2018) notes that "it's hard enough for their sensors to distinguish vehicle exhaust from a solid wall, let alone a billionaire from a homeless person. Right now, developers are focused on more elemental issues, like training the tech to distinguish a human on a bicycle from a parked car, or a car in motion" (2018, para. 9). It is not difficult to imagine, however, a near future in which these problems have been solved and we must then turn to the harder questions of ethics in these sorts of "no-win" driving scenarios.

The autonomous vehicle scenario poses interesting moral questions regarding the relative value of passengers, pedestrians, and other motorists, and illustrates that some of the most challenging ethical questions are perhaps those most relevant to digital environments. Identifying these emerging issues *as ethical* is a key step in the process of becoming morally and digitally literate —in *doing* digital ethics.

Digital Literacy

Concerns like these are part of two overlapping literacies, both necessary for understanding digital ethics. Digital literacy, the process of coming to understand and engage the technologies and information flows that surround us, is one of these. Ethical literacy, or becoming sensitive to, reasoning about, and being motivated to act on emergent ethical issues, is the other. As we see, digital ethics is particularly interesting since it exemplifies the ways in which *epistemic* concerns (about the things and ways we know) are coupled to *ethical* concerns (about the things and ways we value).

Digital ethics is incomplete without an understanding of literacy in digital contexts. Such literacy, like literacy understood in general, is a prerequisite to the ability to understand, to evaluate, and to act on moral problems in digital ethics. Just as one cannot understand a legal document without being able to read and comprehend the words on the pages, it is also true that even more nuanced and advanced comprehension requires some knowledge of the law. In the same way, a person who attempts to understand digital ethics may do so in the most complete way only by knowing and understanding the nature and structures of digital technology as well as the nature and analysis of moral reasoning, theories, and argumentation. Simply put, we cannot understand digital ethics without literacy about both the technical aspects and the ethical aspects of the digital.

"Digital literacy," like most concepts with which we work, has subtleties of meaning that differ among instances of its use. Lankshear and Knobel's collection of essays is titled *Digital Litera*cies (2008) [emphasis ours] in order to make this point explicit. However, as those authors thoughtfully describe (Lankshear & Knobel, 2008), a core idea of digital literacy is that there is an important difference—and relation—between the conceptual and the procedural. Conceptual issues deal with our understanding of the digital, in terms of both its technologies, its environments, and its flows of information. When we think conceptually, we think in terms of ideas and their relations. Much of the work of this book—and of philosophical inquiry more generally—is conceptual work. Procedural issues are those that govern the use, in terms of "tasks, performances, and demonstrations," (Lankshear & Knobel, 2008, p. 3) of digital technologies and their practical applications. When we think procedurally, we think in terms of the mechanics of actions and their relations. This difference between the conceptual and the procedural is an important one. But the more interesting discussion, perhaps, is the relationship between these two.

For example, one might think that the conceptual is necessary for the procedural (and, conversely, the procedural sufficient for the conceptual). This relationship entails that you cannot have procedural literacy without conceptual literacy, but that procedural literacy requires much more than *just* conceptual knowledge. Otherwise put, the sort of procedural knowledge the programmer might have, for instance, does not amount to literacy of procedure unless there is also present conceptual knowledge about structures that constitute the proce-dure. Conversely, procedural literacy is sufficient for conceptual literacy. That is, if you are procedurally literate, it means you have the conceptual literacy on which it stands. But the sort of knowledge the philosopher has about program-ming does not make the philosopher procedurally literate. You might think that this view privileges the practitioner unwarrantedly. However, an example may make this clearer. To be considered "literate," the plumber cannot simply know how to connect parts together to get your septic system to function, but must also understand the nature of the parts, how those systems and parts developed, and the possibilities of their relations. To extend that analogy, the professional development from apprentice to journeyman to master plumber is the process of becoming literate in the art of plumbing—and the master plumber (or perhaps the best among the master plumbers) and *only* the master plumber is conceptually literate (in plumbing) because the master plumber is procedurally literate.

Alternatively, one might think that conceptual literacy is sufficient for pro-cedural literacy (and, conversely, the procedural necessary for the conceptual). Logically, this relationship entails that if you have conceptual literacy about X, you have procedural literacy about X. So, the analytic reasoning historically perceived as the work of professional philosophy is merely armchair philosophy if we take it to be distinct from the procedural literacy. You might think that this relationship privileges the philosopher unwarrantedly: that only the philosopher (the conceptually literate) is procedurally literate, and that sufficient knowledge of procedure is contingent on that conceptual knowledge.

Consider the remaining logical relation between the objects of our current inquiry, conceptual literacy and procedural literacy: that conceptual literacy is necessary *and* sufficient for the procedural. The entailment here is that when-ever you have conceptual literacy, you also have procedural literacy. There is no relevant difference between these two since you cannot have one without the other. This explanation stands in contradistinction to the ways we are all trained as experts. Expertise, in the contemporary global world of academic and professional training, demands that the individual specialize in increasingly nar-row ways. This specialization shuts out possibilities for literacy of the sort that a necessary and sufficient condition entails. So, on such a view, even if you think that indeed it is the case that literacy requires both conceptual and procedural expertise, a pragmatic interpretation (one thinking that the true view is the view that works) denies that option.

A broader lingering worry might be whether literacy itself is a category too broad for anyone to reasonably meet in the first place. If you share in that worry that the demands of literacy cannot practically be met, you are not alone. Those same worries have been ongoing since the sixth century BC Greece, the root of the sort of western philosophical thinking we are doing right now. Plato's conception of the formal world and, by analogous extension, the philosopher-king (Plato, Book VI), stands as a framing concept around which to orient our thinking about the world. The thing to keep in mind in this discussion about the nature of digital literacy is just that relationship between the ideal and the practical.

To overcome the concern about impracticality, consider first that literacy is the sort of thing that develops over time. Becoming literate in the ways we normally conceive of that context, like reading literacy, is a process. Normally functioning (see Norman Daniels' 2000 work in bioethics for a thoughtful definition of this concept) human animals first learn character individuation and pattern recognition before connecting those patterns to aural verbalization leading toward comprehension. And this process develops in complexity with continued training and habituation. This process of becoming literate is reflected in procedural and conceptual literacy as well. So, no one becomes literate in anything all at once.

Second, since literacy is developmental, it is also scalar. The process of becoming an expert is contingent on the nature of specialization and the scope of the knowledge and procedures in the area of expertise. Perhaps the criteria for literacy, as we have laid them out so far, are indeed impossible because they are impractical. Literature on the nature of interdisciplinary, transdisciplinary, and multidisciplinary research pose this very same problem of multi-literacies (e.g., Klein, 2008): Is it possible to be literate in multiple areas of expertise?

Consider the following as a relevant thought experiment: When we are trying to think through the nature and possibilities of artificial intelligence (AI) systems, we must understand how AI algorithms are built. And we also must have some structures of thinking close to hand that guide our understanding of how human intelligence systems are built. Is the human mind merely a complex computational system built of neural networks in the brain? Or is the mind some emergent feature of phenomenal experience as a result of neural networks in the brain? It seems like one can work on AI systems with mere procedural literacy but cannot evaluate when they have artificial intelligence without conceptual literacy. And it seems like one could understand when an artificial system is intelligent with mere conceptual knowledge but cannot effectively develop such a system without procedural knowledge. Thus, conceptual knowledge is necessary but not sufficient for procedural literacy. Literate work in AI requires the shared expertise of both conceptual and procedural knowledge.

In our view, digital literacy is the sort of thing that requires multidisciplinary expertise of both conceptual and procedural knowledges. This high bar is best met (as "multidisciplinary" suggests) by collaborative work among individuals who have diverse expertise. The collaborative philosopher, for example, privileges conceptual knowledge while recognizing the importance of procedural knowledge. And the collaborative procedural expert privileges procedural knowledge while recognizing the importance of conceptual work. The former has some procedural knowledge, and the latter has some conceptual knowledge, the admixture of which is what enables collaboration to happen. In many ways, our authorship of this very book is an example of such collaboration: We are each contributing based on our diverse areas of expertise or development of procedures (building case studies, problematizing concepts, analyzing contemporary issues, critically analyzing different aspects of digital technologies and their underlying software systems, et cetera).

The story we have told here about the nature and possibilities of digital literacy parallels how we view *ethical literacy*, or the process of identifying ethical issues, reasoning about them, and becoming motivated to act upon them in informed ways.

Ethical Literacy

Thinking about the ethical implications of digital technologies is certainly nothing new, historically. Applied ethical issues have played out with the emergence of new technologies. The development of nanotechnology and human genetics, as examples, brought with them a focus on ethical, legal, and social implications (ELSI) that allowed for a conversation about the relative merits of precautionary or proactionary stances toward those technologies. Similarly, approaches to thinking about cyberethics or computer ethics generally, or about robot ethics or AI ethics more specifically, are reminders that emerging technologies bring with them ethical implications.

This kind of work in the ethics of digital technology has taken several forms since the early 2000s. Computer ethics, a focus of philosopher Deborah Johnson's work dating back to 2004, was defined then as "the field that examines ethical issues distinctive to 'an information society'" (Johnson, 2004, p. 69), or a society wherein "human activity and social institutions" have been shaped by computer and information technologies (Johnson, 2004, p. 69). Cyberethics developed out of this perspective with an eye toward incorporating the growing reach and implications of computational and informational technologies. As two of its early theorists noted, "cyberethics issues are concerned with ethical aspects of information as they relate specifically to networked computing and communications devices" (Spinello & Tavani, 2004, p. 1). More recent work at this broad level of inquiry, like that of Luciano Floridi's information ethics (Floridi, 2013), argues that human activity and social institutions are not only shaped by information technologies but are *constituted* by them. Information ethics further interrogates the ethical implications of the information flows that shape and structure human experience. By taking "the digital" to be an umbrella term covering cyber, computer, and information issues, we want to consider not only the human experience of information technologies, but also see "the digital" as an epoch, with human, nonhuman animal, and environmental ethical implications. We seek an ethical approach that enables consistent and coherent ethical analysis across, in information philosopher Luciano Floridi's terms, "levels of abstraction" (Floridi, 2008).

Specific ethical issues, like those implicated by robotics or artificial intelligence (see, for example, Wallach and Allen's 2010 book *Moral Machines*), are in this context the cases through which we work toward a broader understanding of digital ethics. Indeed, we think, the work of general and specific forms of ethics is reflexive; that is, the work of digital ethics both informs and structures as well as is informed by and is structured by the work of the ethics of artificial intelligence and other specific cases.

Becoming literate in digital ethics is the process of gaining particular skills to help us understand issues as they arise. They help us to recognize when a digital issue we face has ethical weight—that is what we mean by developing "ethics sensitivity." And theories and concepts shape the ways we think through ethical issues in a reasoned and responsible way. Finally, ethical theories and concepts must play some sort of motivating role—for what other end is there to thought than action? These three practices—ethics sensitivity, moral reasoning, and moral motivation—make up ethical literacy.

Ethics sensitivity is a synthesis of moral imagination and the recognition of ethical issues (Callahan, 1980). Moral imagination is "an ability to perceive a 'moral point of view'" (Clarkeburn, 2002, p. 440). The second ability, recognizing ethical issues is the "attempt to analyse what has been seen, to recognise the value of moral aspects in a particular situation…[or] to be aware of the moral categories, of the aspects that can be classified as moral and to be able to evaluate their importance to a particular situation" (Clarkeburn, 2002, p. 441). Philosopher Nancy Tuana articulates ethics sensitivity as involving at least three major components: "(1) the ability to determine whether or not a situation involves ethical issues; (2) awareness of the moral intensity of the ethical situation; and (3) the ability to identify the moral virtues or values underlying an ethical situation" (Tuana, 2007, p. 366). The skill of ethics sensitivity enables recognition that a "blend of affective and rational processes" (Tuana, 2007, p. 375) is at work, connecting the agent to the problem and persons involved. The affective, or empathetic, component is vital; for, even if an agent could identify and reason through an ethical issue, that does not entail that "they experience the action as ethical or feel any personal investment in the situation or in trying to respond ethically" (Tuana, 2007, p. 375). One can know what the right thing to do is, and why it is the right thing, but fail to do it. Our focus on empathy later in this book (in Chapter 7) explores this vital connection between affect and reason directly.

The ability to determine whether a situation involves an ethical issue and act rightly in response is of vital importance to digital ethics. The complex interactions of individuals, each with competing and overlapping value priorities and interests, within digital communities (mediated by digital interfaces and technologies) can lead us to believe that one person's ethical issue is another person's personal choice. This is the trap of ethical relativism—the belief that the complexity of epistemic and ethical information denotes an ethical emptiness in which all value beliefs are all and only deeply personal. We wager that you are not an ethical relativist. Instead, we wager that most of us believe in some shared values and ethical commitments, even if the way they work out in the world is complicated and confusing. Ethical sensitivity also attunes us to the moral intensity, or ethical weight, of a particular issue and encourages the practice of naming virtues or values that are shared or in conflict. Those practices of naming give points of shared access for understanding digital ethics.

To offer a hypothetical but all too real example of this value confusion, imagine a self-described "incel" (the self-assigned abbreviation for "involuntarily celibate" individuals (usually men) who hold morally abhorrent views toward the opposite sex) whose post in a Reddit discussion forum is banned by the community moderators. This individual feels like they have been oppressed, their equally valid views suppressed by force. Yet through their action, the community has expressed value commitments: Hateful sex-directed speech is morally abhorrent. In the vast and fast world of the digital, our incel can likely surround himself with a "community" who evidences shared commitment to his values: He can feel part of a community, even if—in reality—that community is marginal. Ethics sensitivity helps us each understand larger shared value contexts and the relative merit of our individual beliefs in them. Our incel has failed to be ethics sensitive, and to identify epistemic and ethical limitations to his views about women. Digital information masks these limitations through curation of communities, making the work of becoming "ethics sensitive" more difficult.

Ethical reasoning is a second key component of ethical literacy, and one of particular importance to philosophical ethics which privileges the skills of thinking critically about ethical issues. Traditional normative theories like deontology, forms of utilitarianism, ethics of care, and virtue ethics (which we will say more about in the next chapter) give us tools by which to reason about the consequences of our actions, our moral responsibility, and the things that morally matter. It gives us the "ability to assess what is held to be valuable in a context" (Tuana, 2007, p. 374)—which some ethicists have identified as the "stakeholders" of an ethical issue. Reasoning about issues to which we are sensitive is the next and essential step, cultivating skills of critical and careful thinking about how best to approach ethical problems, and why.

Finally, becoming "ethics sensitive" and reasoning about a problem does no work in the world (which is the goal of practical ethics) without the push beyond simply thinking about an ethical issue to actually *engaging* it. We might reason that the incel posting on Reddit has a skewed epistemic stance that has led to unethical value claims—but if we do nothing in response, we fail to course-correct our community's ethical landscape. Having moral purpose, courage, and hope (Tuana, 2014, p.169) is not simply rational or cognitive but is affective, asking us to act ethically in a complex world. Such action requires both motivation and strategy; moral motivation is the focus of Chapter 3 and we discuss digital and ethical activism in Chapter 9.

Digital Stakeholders

We mentioned the concept of ethical stakeholders above. Stakeholders are the individuals or groups of individuals who have a stake in the outcome of an activity, idea, or decision. This concept is importantly problematized in digital ethics, given the scope and complexity of digital information.

Traditionally, stakeholders can be local, such as your spouse or neighbor, or global, such as in communities of children living on the other side of the world.

Some stakeholders are obviously identifiable, such as the person with or from whom a student is cheating on an academic exam, but some are more nuanced, such as the other students in the course who are also impacted by the cheating that has occurred. Stakeholders can also sometimes be counterintuitive; for example, there is research from business ethics that suggests that one's competitors in business can be considered legitimate stakeholders in the same fashion as one would consider more traditional audiences such as employees, customers, shareholders, and suppliers (Spence, Coles, & Harris, 2001). The authors of this research cite prior work arguing that businesses have legitimate ethical obligations of accountability that extend to their competitors and their ability to adhere to ethical competitive practices, thus necessitating the inclusion of competitors as stakeholders of business decision making. When we talk about the stakeholders in the digital, we use this term to refer to those individuals impacted by use of digital technologies. This impact can be positive, negative, neutral, or unknown, and the route to that individual may be analog (as in person-to-person) or digital (as in mediated by some digitally networked technology). The degree of the impact also varies widely. Digital technologies can have relatively mild influences on one's life, as when someone is emotionally affected by a Facebook or Twitter post on social media. Or, they can have profound or even life-threating implications, as evidenced in our discussion of autonomous vehicles.

With digital technologies, stakeholders are more difficult to predict and identify than they might be with less ephemeral media. This is because the Internet connects users to massive networks of computers and individuals all over the world, with only some arbitrarily defined rules to separate users from one another. For example, networked videogamers can connect with other users of any demographic from nearly anywhere in the world to collaboratively chat and play. The anonymity of online environments means that these gamers and other individuals using Internet tools may or may not be who they claim to be online. For example, a person claiming to be a man may actually be a woman, and vice versa. Children may actually be adults, and vice versa. And any "person" may truly be a bot, or a piece of software masquerading as a human being. Further, the boundaries between public and private discourse are more blurred than ever before. Social media tools allow individuals to share aspects of their lives and view aspects of others' lives, both business and personal. These other users may range from close family members to casual acquaintances to complete strangers. Aspects of shared lives include everything from academic essays downloaded for cheap from morally dubious websites to personal photos and financial information. Given these complexities, then, how can we conceptualize stakeholders in the digital age?

One method for doing this is to categorize potential stakeholders into broad groups, such as *known* and *unknown* or *local* and *global*. Known stakeholders might be the followers of your Twitter feed, for example, while unknown stakeholders might be those audiences who eventually read your thoughts via retweeting, screen captures, or reading about the tweets in an online news article. Local stakeholders would include those people within your immediate vicinity who are impacted by a technology's use, such as the family members who are deprived of conversation when a child is using an iPad during dinner, or the same child and her peers who benefit from that iPad during math class by using Khan Academy videos and tutorials. Digital stakeholders may also include differentiations among human agents, artificial agential systems, and hybrid-systems in which human agents act through digital interfaces (we think here both digital avatars as well as human-remotely controlled drones). Stakeholders can be identified and engage in digital contexts and explain aspects of digital environments that shape how those agents are evaluated and perceived.

Regardless of how one classifies stakeholders, the point is for designers and developers to be thoughtful about the human beings who are impacted by digital technology as it is being used. Further, as a consumer or user of such technologies, it is advantageous for you to develop an understanding of the ethical implications of the activities you undertake that rely upon digital tools. Not only is this strategy useful to becoming a more informed and critically literate consumer of media and technology, but it is also often strategically advantageous to forming or supporting new relationships with other individuals located across the world.

Why the Nature of the Digital Matters

Perhaps especially in the context of applied ethical issues like the one in which we're currently engaged, the object of thinking *must* be action—we want to enable ourselves and others to guide the policies and practices that have ethical importance. We believe this to be so important that we have dedicated a chapter to it—Chapter 9—later in the book. Of course, all this important talk of theories, arguments, and concepts is merely the staging upon which digital ethics is built. Constructing an understanding of digital ethics requires much more material—and a little *deconstruction* too. So, the next project of this chapter is to tell the story of what we mean by "digital." This work will help us more carefully define the *epistemic* context of digital ethics and perform the work of digital literacy. Drawing on contemporary literature in the ethics of digital and information technologies, we critically address the metaethical foundations that guide the normative commitments of digital ethics.

What scholars, theoreticians, and practitioners have meant by the term "digital" extends across countless concepts and frameworks. One could get lost in an enumeration of those usages. We certainly don't want to lead you down that particular rabbit hole here. But we do want to offer you some perspective on this concept of "the digital" since it informs the work we are here to do in unpacking digital ethics. The first thing to note is that, in normal parlance, "digital" is taken as an adjective describing some other thing: digital currency (like Bitcoin), digital information (as opposed to analog information), digital platforms (like Facebook or Google), or digital technology (like the iPhone), for example(s).

And we could well go on with such a list: digital age, digital media, digital arti-facts, digital objects, digital humanities, et cetera ad nauseam. In much of the work done to date on "digital ethics," the emphasis has been on digital *technolo-gies*. We focus on more fundamental properties of the digital itself, and then think about the implications of the digital on the way it becomes encoded, including in the forms of digital information, technologies, software, and media types.

So, then, what is "the digital" in digital ethics? At a basic level, "digital" refers to a method of data encoding that is differentiated from "analog" meth-odologies. At this level, the digital is a correlative of traditional western phil-osophical binary worldviews, a driving force of thought since Plato. Plato's ontology is dualistic, differentiating a world of forms from our world of rep-resentations. The medieval reconciliation of this view and Catholic dogma maintained this separation of an ideal spiritual world from a non-ideal material one. The very same binary distinction shows up in the empiricism of René Descartes, famous for his mind/body dualism, and in the theologically driven theorizing of Gottfried Wilhelm Leibniz. It shapes the semantic theorizing of Ferdinand de Saussure in his signified/signifier binary and stands as the object of critique offered by French poststructuralists like Derrida and Deleuze, among many others.

But it was Leibniz, in particular, who drove the relation between the binary and the digital in his mathematical work on binary code. His 1703 essay, "Explication de l'Arithmetique Binaire" (Explanation of the Binary Arithmetic, in English), argues for a 0/1 system of arithmetic rather than a system of tens. He argues there that this reduces numbers to "their simplest principles" and demonstrates "a wonderful order" (Leibniz, 2007/1703, p. 1). Leibniz's enthusiasm here is driven not only by the potential of the binary system to order and advance geometrical and arithmetic practices, but also by its historical connections. "What is amaz-ing in this reckoning," he writes, "is that this arithmetic by 0 and 1 is found to contain the mystery of the lines of an ancient King and philosopher named Fuxi, who is believed to have lived more than 4,000 years ago, and whom the Chinese regard as the founder of their empire and their sciences" (2007/1703, p. 2). In ancient Chinese thought, Leibniz found the roots of binary encoding. But he could not have foreseen the incredible impact binary encoding would have on our world of the 21st century.

While the digital as binary-encoding-system has important historical and conceptual connections to philosophical thinking about the world, contempo-rary philosophers like Andrew Galloway and Luciano Floridi, to whom we turn later in this chapter, are driven by a desire to think about the nature of the digital as it has developed into a cultural and technological force in the late 20th and early 21st centuries. The developing expanse of philosophical conceptual work on "the digital" points to the relative youth of this new conversation, tak-ing place in synchrony with the explosive growth of the digital, digital texts (broadly considered), and digital technologies. Our task is not to offer you an analysis of the necessary and sufficient conditions of the digital. But we find that all the various perspectives on the digital share some key common conceptual ground. It is on that ground that we will focus to shape our discussion about digital ethics.

First, the digital fosters *distributedness*, in the form of the reproducibility and transferability of digital information. The digital, embodied in its various technologies, enables the translation of analog originals into binary form, as opposed to mere analogous replicas. Think here of the relationship between the Gutenberg printing press and the copy/paste command on your digital device: In the case of the former, one "original" pattern of metal moveable type is inked and pressed to transfer that pattern to a new medium, like cloth or paper, thereby creating a copy. But this copy is importantly different from the original in medium and form with only the pattern remaining the same. It is the same meaningful information in different form, or what theorist Jean Baudrillard (1981) would call a simulation. Charles Ess (2009) notes that this model of the simulation is radically challenged by the digital. *Digital* copies can be exact replicas of binary information (bracketing questions about lossy compression between file types and differences between analog/physical platforms upon which the transfer and access to digital information depend). The copy/paste command replicates the information at all levels, creating an exact replica (at least intra-programmatically). This blurring of the line between the original and the copy is what led Baudrillard to posit the simulacra, the copy without original. For Ess, this binary replication enables media and the information they contain to converge in the same form and the same location. Your iPhone, for example, lets you browse *a version* of the internet, listen to music, read books like this one, swipe right on Tinder (well! hello, there!), or video chat with friends or that creepy guy living in his mom's basement. This convergence of information on a single device is made possible because the fundamental structure of digital information is compatible with a single platform in a way that previous forms of that same information have not been. This idea of distributedness marks one fundamental difference between the analog and the digital.

Second, the digital "greases" information in its *procedurality*. This mechanical metaphor of grease is taken up from theorist James Moor, who developed it in a 1997 article on privacy in the information age, writing "when information is computerized, it is *greased* to slide easily and quickly to many ports of call" (Moor, 1997, p. 27). The digital's procedurality opens the ability to transport, copy, and distribute information efficiently and effectively between sources, owners, and storage locations. Not too long ago, researchers spent substantial time in library stacks, finding printed sources of information, and transcribing some of that information onto physical note cards. They would then collate those notecards into a physical outline of ideas before typing out an argument and research article to be saved on a digital device (… or analog device). The rise of the digital has greased the flow of that information, allowing researchers to access digital versions of those same sources, create digital copies, type out digital notes, and transcribe into digital form—avoiding most if not all changes of medium during the research process. This is a morally and practically significant point, since it makes practices like plagiarism so much easier than it used to be in that it is facilitated by digital technologies.

Third, the digital is *embedded*. Unlike analog forms of information and technologies, their digital form is unavoidably pervasive. Binary digital data is easily embedded into new and different information streams, news feeds, and software APIs in ways not possible for analog data. Thus the digital pervades the ways we work, think, consume, understand, and engage in almost every facet of our daily interactions. It continues to become inextricably linked to us through our technologies and within our environments. This embeddedness has important ethical implications for problems related to privacy, autonomy, agency, and empathy.

These foundational characteristics of the digital (distributedness, procedurality, and embeddedness) take other forms in the various existing literatures around digital ethics. For example, Yoo, Henfridsson, and Lyytinen (2010) argue that digital technologies point to the homogenization of data (content as separate from the medium), reprogrammability (the same device can perform a wide range of functions), and self-referentiality (digital innovations requires the use of digital technologies) (p. 726). The concepts that other scholars have identified do similar work: Ess' (2009) convergence aligning with Yoo's "homogenization," and Moor's (1997) greasedness aligning with Yoo's "reprogrammability." The approach to identifying a cluster of concepts orients ethicists to problems in particular ways. The three we have identified we think are fundamental to the digital, as opposed to its technologies or software; for example, the self-referentiality of digital devices is made possible because of these properties of the digital itself.

Consider our previous examples of digital texts (copied and exchanged): that story can be read as a closing of the space for analog media, like physical texts. Instead, the distributedness, procedurality, and embeddedness of the digital lead to the development of new digital tools to create new digital texts and other media representations of information, all digital. And, of course, even given the remarkable progression of the speed and scope of digital information, analog texts continue to survive, too, despite continuing predictions of their demise. The digital exists in relation to the analog.

But we should say a little more about the open landscape in which we are working here. As we mention above, scholars and practitioners have and will continue to engage in working out the very idea of "the digital," adding more to its description including but not limited to concepts like editability, interactivity, reprogrammability, distributedness (Kallinikos, Aaltonen, & Marton, 2013) and "multiple inheritances" in distributed settings, meaning there is no single owner that owns the platform core and dictates its design hierarchy (Henfridsson & Bygstad, 2013). Consider popular cultural theorist Alexander Galloway's thinking on the digital. Galloway, in a 2012 lecture and then a 2014 book, works out a difference between "flat" and "deep" digitality, where flat digitality is from the "multiplexing of the object," and deep digitality is from the "multiplexing of the subject." Elsewhere, he describes flat digitality as analogous to grids of pixels or "windows" in a computer desktop interface—grids of cells working together to create a whole. And he argues that deep digitality might be analogous to a multiplicity of points of view (Galloway, 2014, pp. 68–69). Galloway's thinking on the topic is deeply informed by his reading of philosopher Gilles Deleuze (though Galloway himself has noted that cultural theorists' reliance on Deleuzian concepts is "getting out of hand" (Galloway, 2017)) and by the anti-philosophy of French thinker Francois Laruelle. Galloway's characterization of the digital is another addition to the already complex landscape of conceptualizations of the digital, reaching from the quite specific to the remarkably broad. In his 2014 book Galloway argues that "digital," on his broadest read, "is the basic distinction that makes it possible to make any distinction at all. The digital is the capacity to divide things and make distinctions between them" (Galloway, 2014: xxix introduction).

This broad definition brings up the last perspective we want to address on this issue of defining the digital. Some theorists, like Galloway and philosopher of information Luciano Floridi (see Floridi, 2011), have begun to articulate a view that "the digital" is not some historical moment or cultural shift but, instead, a representational model of the way the world *actually is*. The digital as making distinctions possible is a view or infrastructure of thought about the digital that governs how the world is understood as constituted all the way down. For Galloway and Floridi, the world is fundamentally one of distinctions, differences, made possible by flows of information. And the thing that makes the digital a particularly relevant topic of inquiry *now* is not that it is particularly novel, but that particular characteristics of it are radically divergent from past forms. One way to view this radical divergence is in terms of the *speed and scale* of information flow. As technologies have developed, they have enabled exponential increases in the volume, variety, and velocity (the "3Vs" (Laney, 2001)) of information exchange. This is the so-called information revolution, the breath-taking pace and unprecedented scope of acceleration of information, and is something we discuss in more detail later, particularly in Chapter 8 where we evaluate the impact of information's speed on free choice. In this view, the whole history of human technological development is the history of digitization, the increased speed and scope of information exchange. And one only need consider the similarly astonishing speed and scale of human-caused climate change to begin to see an argument for thinking about *the digital* as defined sufficiently by concepts of scale and speed, applying not only to information technologies but also to geographic change, social systems, and biological functioning. The digital is the recognition of the governing role ontological infrastructures like these play in describing the world.

Understanding digital ethics requires this kind of epistemic work; that is, it requires the work of understanding *the digital* as much as understanding ethics. The nature of the digital is coupled to the ethical implications of its technologies through the ways by which information flows challenge and change patterns of knowledge. This kind of epistemic-ethical coupling is not distinct to digital ethics but is essential to it.

Next Up

In the next chapter, we examine the second component of ethical literacy, the ability to reason about ethical issues. We outline the basics of the "big" normative theories in philosophical ethics and argue for a pragmatic pluralism in reasoning ethically. A series of case studies and examples serve to demonstrate how much context matters to the ways reasoning applies.

References

Baudrillard, J. (1994[1981]). *Simulation and simulacra.* (S. F. Glaser, Trans.) Ann Arbor, MI: University of Michigan Press.

Bonnefon, J.-F., Shariff, A., & Rahwan, I. (2016). The social dilemma of autonomous vehicles. *Science, 352*(6293), 1573–1576.

Boorstin, J. (2016). Facebook tackles fake news. *CNBC.* Retrieved from https://www.cnbc.com/2016/12/15/facebook-tackles-fake-news.html.

Callahan, D. (1980). Goals in the teaching of ethics, In: D. Callahan (Ed.), *Ethics teaching in higher education* (pp. 61–80). New York: Plenum.

Clarkeburn, H. (2002). A test for ethical sensitivity in science. *Journal of Moral Education, 31*(4), 440–441.

Constine, Josh. (2016). Zuckerberg implies Facebook is a media company, just "Not a traditional media company. *TechCrunch.* Retrieved from https://techcrunch.com/2016/12/21/fbonc/.

Daniels, N. (2000). Normal functioning and the treatment-enhancement distinction. *Cambridge Quarterly of Healthcare Ethics, 9,* 309–322.

Ess, C. (2009). *Digital media ethics.* Cambridge: Polity Press.

Floridi, L. (2008). The method of levels of abstraction. *Minds and Machines, 18*(3), 303–329.

Floridi, L. (2011). *The philosophy of information.* Cambridge: Oxford University Press.

Floridi, L. (2013). *The ethics of information.* Cambridge: Oxford University Press.

Francis, Diane M. (2017). Facebook's ethical problem. *HuffPost.com.* Retrieved from https://www.huffpost.com/entry/facebooks-ethical-problem_b_59c923f9e4b0b7022a646c61.

Galloway, A. (2012, November). 10 Theses on the digital. [Vimeo video]. Retrieved from https://vimeo.com/48727142.

Galloway, A. (2014). *Laruelle: Against the digital.* Minneapolis, MN: University of Minnesota Press.

Galloway, A. (2017, November 5). Peak Deleuze and the red bull sublime. Retrieved from http://cultureandcommunication.org/galloway/peak-deleuze-and-the-red-bull-sublime.

Gere, C. (2008). *Digital culture.* London: Reaktion Books.

Henfridsson, O., & Bygstad, B. (2013). The generative mechanisms of digital infrastructure evolution. *MIS Quarterly, 37*(3), 907–931.

Hill, B. (2016, June 24). Self-driving cars will likely have to deal with the harsh reality of who lives and who dies. *Hothardware.com.* Retrieved from https://hothardware.com/news/self-driving-cars-will-likely-have-to-deal-with-the-harsh-reality-of-who-lives-and-who-dies.

Johnson, D. G. (2004). Computer ethics. In L. Floridi (Ed.), *The Blackwell guide to the philosophy of computing and information* (pp. 65–75), New York: Blackwell.

Kallinikos, J., Aaltonen, A., & Marton, A. (2013). The ambivalent ontology of digital artifacts. *MIS Quarterly, 37*(2), 357–370.

Klein, J.T. (2008). Evaluation of interdisciplinary and transdisciplinary research: a literature review. *American Journal of Preventive Medicine 35*(2S), S116–S123.

Laney, D. (2001). 3D data management: controlling data volume, variety, and velocity. *MetaGroup 949.*

Lankshear, C. & Knobel, M. (Eds.). (2008). Introduction: Digital literacies – concepts, policies, and practices. In *Digital literacies: Concepts, policies, and practices* (pp. 1–13). New York: Peter Lang.

Leibniz, G.W. (2007). Explanation of binary arithmetic. (L. Strickland, Trans.). Originally published in *Die Mathematische Schriften von Gottfried Wilhelm Leibniz, vol. VII.* C.I. Gerhardt (Ed.). pp. 223–227. Retrieved from http://www.leibniz-translations. com/pdf/binary.pdf.

Manovich, L. (2013). *Software takes command.* New York: Bloomsbury.

Marshall, A. (2018, October 24). What can the trolley problem teach self-driving car engineers? *Wired.* Retrieved from https://www.wired.com/story/trolley-problem-teach-self-driving-car-engineers/.

Maxmen, A. (2018, October 24). Self-driving car dilemmas reveal that moral choices are not universal: Survey maps global variations in ethics for programming autonomous vehicles. *Nature.* Retrieved from https://www.nature.com/articles/d41586-018-07135-0.

MIT Media Lab. (n.d.). Moral machine: Human perspectives on machine ethics. Retrieved from http://moralmachine.mit.edu/.

Moor, J. (1997). Towards a theory of privacy in the information age. *Computers and Society, 27*(3), 27–32.

Plato. (360 B.C.E.). *The republic.* B. Jowett, Trans. The Internet Classics Archive. Retrieved October 1, 2018 from http://classics.mit.edu/Plato/republic.7.vi.html.

Schroeder, S. (2019). "Facebook tackles fake news in the UL with a new fact-checking service. *Mashable.* Retrieved from https://mashable.com/article/facebook-fake-news-uk/.

Spence, L. J., Coles, A.-M., & Harris, L. (2001). The forgotten stakeholder? Ethics and social responsibility in relation to competitors. *Business and Society Review, 106*(4), 331–352.

Spielberg, S. (Director), Molen, G. R., Curtis, B., Parkes, W. F., de Bont, J. (Producers), Frank, S., & Cohen, J. (Writers). (2002). *Minority report* [Motion Picture]. United States: Twentieth Century-Fox.

Spinello, R. A. & Tavani, H.T. (Eds.). (2004). Introduction to chapter 1: Cybertechnologies, ethical concepts, and methodological frameworks: An introduction to cyberethics. In *Readings in Cyberethics.* Boston, MA: Jones and Bartlett Publishers.

Swisher, K. & Wagner, K. (2018). Mark Zuckerberg says he's 'open' to testifying to Congress, fixes will cost 'many millions' and he "feels really bad'. *Recode.* Retrieved from https://www.recode.net/2018/3/21/17149964/facebook-ceo-mark-zuckerberg -congress-data-privacy-cambridge-analytica.

Tuana, N. (2007). Conceptualizing moral literacy. *Journal of Education Administration, 45*(4), 364–378.

Tuana, N. (2014). Being affected by climate change. In K. Shockley and A. Light (Eds.), *Ethics and the Anthropocene* (forthcoming). Cambridge, MA: MIT Press.

Thomson, J. J. (1985). The trolley problem. *The Yale Law Journal, 94*(6), 1395–1415.

Wallach, W. & Allen, C. (2010). *Moral machines: Teaching robots right from wrong.* New York: Oxford University Press.

Worland, J. (2017, NovemberDecember). Why self-driving cars might not lead to a huge drop in fuel consumption. *Time, 190*(2223), 30.

Yoo, Y., Henfridsson, O., & Lyytinen, K. (2010). The new organizing logic of digital innovation: An agenda for information systems research. *Information Systems Research, 21*(4), 724–735.

2
MORAL VIEWPOINTS IN DIGITAL CONTEXTS

Ethical reasoning is more complex than rules or lower-level procedures for behavior, like algorithms. Algorithms, rules, and procedures (discussed in more detail in Chapter 5) are at best simply minimal sets of guidelines and not the same as providing reasons and justifications for them. While guidelines are useful, they are incomplete for ethical reasoning. "Real-life" ethical issues and dilemmas are dynamic and complex, and so not easily—or not at all—formulaic. It is possible that making the right decision and doing the right thing may be based on guidelines or even on feeling, but in such cases the right or good decision or action has been performed accidentally at best. Especially when decisions we make and actions we perform have important, serious consequences on ourselves or others, we should not leave the consequences of our actions to feelings or to unexamined procedures. Using facts, principles, and reasoning rather than feeling, intuition, or simple processes offers us a way in which consensus building for careful decisions and actions is more likely ensured.

There is nothing inherently objectionable about lists of rules and procedures or the way we feel about things, but neither procedures nor emotions make things either good or bad or right or wrong. Take the following example, which uses the "yuk factor" (Midgley, 2000) to illustrate the point. On YouTube, there are videos showing the processes involved in knee and hip replacement surgeries. Surgeons use mallets and drills to remove natural tissues and to insert new, artificial ones. The procedure is graphic, bloody, and violent-looking. Some people find the videos disgusting and unpleasant to watch, yet it does not mean that we ought to discontinue performing knee and hip replacements. On the other hand, someone may be motivated by the "yuk factor" to devise different ways to achieve the same or better surgical results, minimizing pain as the result of sympathetic or empathetic feelings for other people and animals. In this example, both feeling and "procedures" combine to create better conditions than had previously obtained.

Finding solutions to moral questions and associated decisions and actions are complex, variable, robust, and dynamic. One of our goals in this book is to illustrate and to explain the complexity, variety, and difficulty of moral issues and decision making and actions associated with them. We argue for a means of consensus building through dialogue that results in agreement even when feelings, procedures, and theories seem in conflict. We do not, therefore, argue for a specific ethical theory or for hard and fast rules. We argue for a pluralistic and pragmatic method of reasoning about ethics that is amenable to the use of various theoretical points of view. Using a pluralistic and pragmatic approach to analyzing moral problems may not always lead to agreement on some issues, but in using it, it is possible for us to delve deeply into the core of issues to gain as much information and use the best reasoning we have available to reach a consensus that will, in turn, lead to policies, procedures, rules, and decisions that are the best we can make. We contend that consensus building, dialogue, and reasoned argumentation may lead us to decisions and actions that are well-considered and thoughtful rather than rash, inappropriately emotionally charged, and that will be the best decisions about a case we are able to derive. There are cases, especially complex and important ones, for which definitive answers cannot be found easily—or at all. It is through case studies in digital ethics that we hope to foster an educational environment for readers of this book that helps them to employ good reasoning and make good decisions about issues involving the digital world.

CASE STUDY: TEXT-ASSISTED SUICIDE?

In a text message, a 17-year-old young woman, Michelle Carter from Massachusetts, urged her boyfriend, Conrad Roy III, to kill himself. Mr. Roy had decided to kill himself by asphyxiation from carbon monoxide using his vehicle's exhaust. Apparently having second thoughts, he contacted Ms. Carter, who told him in a text message to "get back in" the vehicle. Mr. Roy died after re-entering the vehicle. Subsequently, Michelle Carter was convicted in June, 2017, of involuntary manslaughter in a Massachusetts court. In a *Newsweek* article online, David Rossman (2017) wrote that Roy's death is a tragedy, an "atrocious and cold-hearted action," and morally contemptible "by almost any ethical standard." But are these simple, straightforward answers to pre-conceived questions so simple, and are they the correct answers? Let us examine the reporter's commentary as an introduction to the complexities of cases with moral import in digital ethics.

It is alleged that Michelle Carter told a friend that she "told him [Conrad Roy] to get back in ... because I knew he would do it all over again the next day and I couldn't have him live the way he was living anymore. I couldn't do it, I wouldn't let him." The reporter claims that Ms. Carter's action is "morally contemptible ... [b]y almost any ethical standard." To what ethical standards is the reporter appealing? This question may be the most important of all those the reporter asked in the article because the term "ethical standard" itself is unclear, and simply to assert that her actions are contemptible by "almost any ethical standard" is vacuous at best, and surely misleading—especially to those who are both unfamiliar with ethical theories in general and who may believe that their own ethical standard is the only one to which it is worth appealing.

Would we have questions concerning what Carter said to her boyfriend if their conversation had been in person or on an unmonitored or unrecorded phone line? No one except Ms. Carter and Mr. Roy would have known the contents of the conversation, and what she might have said later to a friend regarding her views of Mr. Roy's life would be plausibly deniable. But what she said to her boyfriend was said using a digital device (a text message on a phone), so it was captured and recorded, making it undeniable that she told him to "get back in." What is relevant for our purposes is the moral evaluation of the actions, intentions, and character of a person who said to a young man having second thoughts about suicide simply to "get back in" the truck—and ultimately to die.

You will see that we leave the discussion of cases presented on digital ethics open-ended. This allows you, our reader, to discuss, consider, deliberate about, and make decisions on your own—building literacy in digital ethics. The titles of cases presented throughout the book are all listed in the Appendix, along with a method for conducting detailed case studies, should readers wish to work them out as exercises. Reading what we think about the cases is not as important as ensuring that you can take a fresh and independent look at the cases and engage more fully in moral reasoning and decision making by being an active participant in case evaluation.

Ethical Theories, Principles, and Problems

We take the approach that there is a three-part process in analyzing moral problems in a reasoned way. They are to (1) identify moral problems and relevant facts, (2) formulate appropriate questions about them, and (3) make decisions regarding options for morally appropriate action through argumentation. We discussed identification as a part of ethics sensitivity in the last chapter, but it is not enough to be aware of problems, to theorize about them or to propose solutions on which one does not intend to act or expect others to act. Making moral judgments is a normative process—that is, in making moral judgments, one is doing more than simply stating an opinion or one's considered judgment on an issue. Making a moral judgment is *prescribing* action for oneself and others based, in an ideal condition, on good and accurate information, good reasoning, and reliable moral principles. It is essential to engage in ethical reflection and carefully considered action to make decisions that are good and right, to perform actions that are appropriate in the digital and social worlds, and to work toward an ethical stance regarding digital technology that improves the lives and circumstances of individuals and societies through positive action for individual and social good. This ongoing work of ethics is a process of becoming literate, both in the digital and moral senses.

A Primer on Ethical Theories

In Western ethics, there are several prominent theories for identifying, raising questions about, proposing solutions to, and acting on moral problems that are as relevant to the problems of digital ethics as they are to any applied ethical context. Knowing the basic elements of these theories will help to conceptualize problems in digital ethics in an inclusive fashion. You may find that it is possible to identify a moral problem embedded in a particular case study that you would not have otherwise noticed if you had not been aware of concepts, principles, and arguments concerning the theoretical basis of ethics. Analyzing moral issues includes making distinctions between types of ethics and not simply between types of ethical theories, and discussing some of the basic problems of theories, their meaning(s), and their applications in particular contexts.

Moral inquiry, or the subject-matter of ethics, is of three kinds. Normative theories are also called "theoretical ethics." Such theories are prescriptive. That is, in being prescriptive, moral theories help us to determine what ought to be the case, what ought to be done, and where moral problems arise in cases. Moral theories are therefore prescriptive in that they tell us, at least in a general way, how to approach and to avoid or to solve moral issues, problems, and dilemmas. Applied ethics is a branch of ethical inquiry concentrated on resolving moral problems of a particular kind. This book is primarily about applied ethics since its content centers on the analysis of ethical issues in scenarios involving or mediated by digital technologies. Other areas of applied ethics are environmental ethics, animal ethics, business ethics, medical ethics, engineering ethics, and law enforcement ethics, among many others. Meta-ethics is an inquiry into the nature of the good and the right and the meanings of moral terms. This is the branch of ethics that may ground the application of ethics to specific contexts. These three kinds of ethics are intimately related: We cannot answer practical questions without structured theories to help us reason, and we cannot reason about things we do not clearly understand. We will continue to argue throughout this book that the meaning of ethical terms is challenged in the digital, and that digital literacy can help us understand these challenges.

Moral theories provide background for complex processes of moral discussion and deliberation that go well beyond applying one theory to a case or problem and making a decision based only on it. While it may be easy to apply one point of view to a problem and consider it "solved," moral problems are messy and usually complicated, multifaceted, and dynamic. So, it is necessary and advantageous in decision making to consider facts, questions, perceptions, various theoretical points of view, and even feelings or emotions. We argue for the use of multiple theories and points of view in the analysis of cases, which results in more complete, thoughtful, and consensus-building analysis of moral problems. The use of research and reason in a full and nuanced, intelligent, and multi-faceted manner will lead to careful and thoughtful decisions about ethical issues. For a discussion of a means by which you may approach the analysis of and solutions to moral problems, see the Appendix of this book. Next in this chapter, we offer a brief presentation of the "big" moral theories as a set of tools for cultivating ethical reasoning skills.

Before we do, remember that ethics is not like mathematics or formal logic. It is not a subject-area in which there are formal methods of solving problems. If this is disappointing to you and it leads you to believe that there are no "answers" to questions and no solutions to problems in ethics, remember that ethics is about values, and values do not lend themselves handily (or at all) to simple calculation. That, however, does not absolve us of the responsibility to seek answers to our moral questions and to employ the best reasoning of which we are capable.

Utilitarianism

Utilitarianism is probably the most simple, easily identifiable, straightforward, and familiar of all major Western ethical traditions. The principle upon which utilitarianism is based is the "Principle of Utility." John Stuart Mill, a 19th century British philosopher, formulated the Principle of Utility such that actions are right insofar as they promote happiness and wrong as they do not do so. For the utilitarian, happiness is pleasure and the absence of pain; unhappiness is pain and the absence of pleasure (Mill, 1863/2001). The most important moral consideration in Utilitarianism is the consequences of actions. Where it is not possible to create happiness or pleasure, the utilitarian advocates the minimization of pain.

Utilitarians focus on the good. That is, utilitarians are consequentialist thinkers and actors, requiring that we take into careful account the effects of actions, policies, procedures, structures, and thoughts put into action on the people, animals, and things involved. The "greatest happiness" is a nuanced and complex feature of utilitarian thinking. It is not a simple matter of considering mere "pleasure," but instead the quality of pleasure that is derived from the individual and social contexts in which our moral lives take place.

Deontology

Deontology may be attractive to those preferring principles advancing the right rather than the good. Immanuel Kant, a German philosopher of the 18th century, formulated deontology in *Grounding for the Metaphysics of Morals* (1785/1993). For the deontologist, right is determined by rationally calculable factors in moral reasoning such that moral decisions cannot be based on mere empirical concerns such as consequences of actions, happiness, and well-being. Such empirical, experiential concerns are subject to vagaries of experience and are therefore too uncertain to be used as determinants of moral rightness. Kant therefore proposed that moral decision making must be based on a universal rule. Unlike utilitarian considerations of the potential results of actions, the deontologist is concerned with the purity of motive or intent relative to the rule derived.

Deontology requires acting on the basis of "the categorical imperative," the ultimate rule of the theory which demands that in determining the right thing to do, one must act on a maxim that can be willed to be universally binding, or to be a universal law (Kant, 1785/1993). In other words, the categorical imperative gives us a decision procedure for action and a guide for action in that one must act always such that humanity is never treated as a means to an end, but always as an end (Kant, 1785/1993). This means that it is never morally appropriate to use any rational being to achieve a desired end, because to do so would be to treat a person (or any rational being) as a "thing." Instead, rational beings must always be treated with respect, as autonomous moral agents, as ends in themselves.

Virtue Ethics

Virtue ethics, founded in ancient Greece in the philosophy of Aristotle in the *Nicomachean Ethics* (350 B.C.E./1999), is a theory in which neither duty nor consequences are the sole determining factors of a moral or good life or a moral or good action. Where deontology and utilitarianism center on the individual and calculating duties or consequences, virtue ethics depends upon complex social dynamics and the concept of a good life.

In virtue ethics, a human being's ultimate goal is to achieve "happiness," characterized as a complete, dignified, fulfilled life. Happiness is easily lost, but it is hard won: It is an activity experienced, lived, and practiced. It does not consist simply in obtaining and keeping material goods or wealth, nor solely in living in a society in which one's needs are met. The human being is happy (that is, the human being lives a good life) when her or his rational capacities are respected and reasonable desires are satisfied and reciprocated between the individual and the community.

Virtues are excellences and the "virtue" of a thing is in its function. When a thing functions well, we call it good. When it does not do so, we call it bad (or even defective). Rationality is that which is both specific to human beings and that which we at least in principle do best; the virtue of a human being is in the proper use of reason. Virtue applies to what a human being does, and when acting rationally, one is virtuous. Aristotle applies rational action to the realm of human virtues and actions in "The Doctrine of the Mean," in which virtue is a variable mean between two vices.

Vices are excesses and deficiencies while virtue is the perfection of a thing. Virtues vary by person, time, place, and case. So, expecting a person who is naturally fearful of crowds to go comfortably to perform onstage in a comedy club would be unreasonable, but the fearful person who does so might be considered courageous for getting on the stage. Alternately, a naturally gregarious person would hardly be considered courageous for simply getting on stage and telling a few jokes. Virtues, as perfections of character and action, are not subject to purely rational evaluation even though the development of virtues is insepara- bly connected to rationality. Virtues are developed through practice and become habitual and thus part of who a person is. Just as one does not become adept at playing first person shooter videogames by simply clicking a mouse or gamepad button, and instead must practice regularly to become excellent at hitting targets and understanding the point of the games, a person does not become generous simply by giving to a charity or performing some other act of generosity once, or even every now and then, but instead cultivates the virtue by practicing it regu- larly. Such practice is accompanied by the characteristics of the community in which one lives, and as we are (according to Aristotle and other virtue ethicists) social or political animals, the virtues are developed in conjunction with the communities of which we are a part. Virtues are a key part of ethical literacy for their role in *moral motivation*, which we address specifically in the next chapter.

Contractarian Ethics

Contractarian ethics has close affinities to the socio-political conception of social contract theory in which adherents contend that our political existence is prop- erly created and governed by agreement. The contractarianism of John Locke in the *Second Treatise of Government* (1689/1980), for example, contended that we form governments and laws for protection against "lions" in our midst, the "nox- ious few" who violate norms and laws, not against every other person. So, the notion that "we the people" wish to create a "more perfect union" is an expres- sion of the contention that people create their governments. It should not be the case that governments are foisted upon the people against their wishes, contrary to their interests, and without consideration of their desires to be governed (or to govern themselves) in a particular way.

It is important to note that Kant was a contract theorist and that his adher- ence to contractarian political thinking is parallel to the way in which he derived his deontological moral conclusions. The reason that the categorical imperative requires respect for persons is intimately associated with the belief that rational human beings are fully capable of giving moral laws to themselves. In other words, rational human beings are autonomous moral agents who, using their own intelligence, are able independently to construct appropriate rules of moral- ity and social life. They do not need to be told or ordered what to do. If human beings are capable of deriving moral rules independently of others, and when reason is applied properly they derive the same moral rule(s), it is also the case that such people, who possess inherent dignity and value, ought to be afforded the opportunity to participate fully in the creation of moral and other rules for social and political systems. For contractarian ethics, morality is constituted by our agreement on at least basic principles to guide our social interactions. We address contractarianism directly later in Chapter 5 and these threads of governance, policy, and moral rules show up regularly throughout the book.

The Ethics of Care

Care ethics is an ethical theory concentrating in part on the limited theoretical range of traditional ethical theories and offering what its adherents consider a richer and more applicable theory. The ethics of care is a very recent development in Western moral thought, deriving largely from analysis of psychological development theories and a transformation in ways of thinking about ethics by many feminist theorists. Carol Gilligan, in her groundbreaking work *In a Different Voice* (1982), developed a new way of thinking about moral development. Her work breaks away from that of her mentor and colleague, Lawrence Kohlberg (e.g., 1981). Kohlberg identified stages of moral development ranging from the very basic and rule/force-based in children to principle-based moral reasoning in adults. With this distinction, Kohlberg concluded that principle-based moral development was the highest level of moral reasoning and action. Gilligan, however, noted a striking fact about Kohlberg's theory of moral development in that women and girls usually did not exhibit in the experimenters' observations the same kind of reasoning or moral orientations as boys and men. More specifically, Gilligan noted that girls and women are more attuned to relationships and are more likely to confer with others (rather than appealing simply and singly to principles) to make moral decisions and judgments. On Kohlberg's scale of moral development, this fact puts women and girls on a lower stage of moral development than men and boys, generally. In Kohlberg's scheme, the highest stage of moral development is much like that of deontology, which is, as you have already seen, completely principle-based and derived from (supposedly) the purest reasoning. Gilligan and feminist ethicists doubt that considering the interests of others, conferring with them, and taking into account our "relational" selves constitute grounds for proclaiming (most) women and girls to possess moral acumen less developed and valuable as that of (most) men and boys.

Much of the distinction that theorists in care ethics wish to make hinges on the way in which traditional moral theories (including contract theory, utilitarianism, deontology, and even virtue ethics) focus on "justice" or other abstractions in ethics. Care ethicists refer to traditional moral theories as "ethics of justice" and contrast such theories with an "ethics of care." What is more, they often make a distinction between ethics used in a public sphere and how it is different from the private sphere. Traditionally speaking, the work of women and the realm of women's influence in society extended not much farther than, and perhaps only as far as, the family and domestic relations in general. Women's work was to care for children and manage household affairs. It was the men or man of a family who went into the public sphere where the "important" work of making social and political rules and procedures took place. But theorists of care ethics astutely point out that what makes possible life in the social and political realms is that children and others are taken care of in families as relational beings by considering their individual and personal interests (taking care of a person is more than simply seeing to it that their basic biological and functional needs are met), and that the moral is not just or specifically social. It is also personal. To ignore this fact is to impoverish ethical theorizing and ethical action by failing to see that our relationships to each other as human beings are much more varied and complex than the derivation of rules and procedures for the proper organization of society would allow. Impoverishment of ethical theorizing also applies to virtue ethics because its dependence on moral exemplars (those persons and institutions to whom an individual looks to develop the virtues) emphasizes compliance and the elevation of the "public" realm of "justice" over the more fluid and developmentally important "private" realm of "care" in which human beings also live.

Theorizing about the ethics of care is not done by providing rules, procedures, and exemplars to follow. Instead, care ethics encourages us to see ourselves as relational beings. As such, moral thought and action include the private as well as the public care, as well as justice. Care necessarily includes considering the specific needs of individuals and groups, and the way to include them is to ensure that they and their interests are part of conversations, procedures, and policies and not that conversations, procedures, and policies take place without them and without considering their points of view. In other words, women, children, the disabled and differently abled, some nonhuman animals, and perhaps even digital agents, who have been regularly excluded from participation in decision-making are, according to a care ethics approach, considered active, engaged, and essential participants. Not to recognize the multifaceted variety of experiences and actions is to present our theories of ethics and our ethical actions as those of isolated, atomistic individuals who have no necessary or actual relations to each other in families, in social groups, and in interests that provide for and necessarily transcend the foundation for abstract theorizing.

Pragmatic Pluralism for Digital Ethics

Elements of virtue ethics and the ethics of care constitute a core of the approach to ethics that we endorse in this book because we take a *pluralistic* approach to ethical theorizing and application. We contend that ethical decision making demands consideration from perspectives of multiple theories of ethics, representing many and varied points of view. For example, later in Chapter 5, you will see how pluralistic ethics can be used as a heuristic model for generating many different ethical considerations and perspectives for machine ethics: It stands as a richer if not more complex approach then developing "moral engines" based on singular specialized theories. Our general commitment to value pluralism throughout this project sustains a more robust, inclusive, and practically applicable ethical decision-making approach than any singular, monolithic moral structure for argument and action can provide. And it is more than a mere practical heuristic to ethical decision making.

Pragmatism is a methodological development in Western philosophy developed by early 20th century American thinker Charles Sanders Peirce and later championed by William James and John Dewey. A way in which to understand our approach in this book is to take into consideration William James's view that "truth happens to an idea" (1907) and that a tendency toward absolutism is, as James put it, a "weakness of our nature from which we ought to free ourselves" (1897). Further, like Peirce, our goal is to bracket the search for Truth since, even while it should orient us to reasoning, it itself is unattainable given the complexity of our situatedness in particular context. In the place of "Truth," we search for *truths* in the form of beliefs we hold that are generally unassailable by doubt at some specific time and for some specific place. Peirce's view was that we "fix" (1877) or establish our beliefs properly through an experimental (or experiential) method rather than simply to adhere dogmatically to a single theory. Such an approach is limiting: while it may lead to an answer to some ethical problem, it ignores perspectives and context which are vital.

Truth, for the pragmatist, grows in communities whose members are all committed to striving toward it. And like John Dewey, we hold that traditional ethical problems such as a search for "The Good" is too abstract for our efforts to solve problems here and now, and that we ought to seek solutions specific for this time and place and context (see Dewey, 1920), even while we remain oriented toward ethical ideals. Beyond this, we also take Richard Rorty's view that in the face of the contingency of our lives in this world, and the variations in people, places, ideals, concerns, problems, issues, and their solutions, we ought to seek solidarity with others in their and our search for solutions to problems (Rorty, 1989). With all this in mind, our pluralistic approach to digital ethics is that there may very well be a variety of ways by which to conceive of and solve moral problems in digital contexts. Such a pragmatic pluralism does not sink us back into relativism, where any agreed-upon solution is a good one; rather, it acknowledges the epistemic and ethical complexity that leaves us oriented toward but unable to directly access Truth with a Big-T.

This talk of pragmatic use of normative theory, and the search for Truth or truths is all contingent on reasoning. Reasoning, that paradigmatic distinctive human capacity, has been the driving force of Western philosophy through to the end of the 20th century at least. From Ancient thought through the Enlightenment, to critical theory and the rise of postmodernism, reason has been taken, often for granted, as the engine that drives the human quest for knowledge and goodness. It has also been critiqued as underutilized, ignored, overwhelmed, and damaged by manipulation. To reason is simply to apply the rules of logic to evaluate claims of knowledge about the world. To reason ethically is to apply reasoning to moral matters. We describe this as a necessary condition of moral literacy, and others have argued before that ethical inquiry is a craft, with students apprentices in that craft (Lipman, 1987). Both views give voice to the same idea that ethical reasoning is a process by which the complexities of a moral problem are made clearer by carefully thinking through them in a structured way. Philosopher Nancy Tuana, in her work in ethical leadership education, has argued that ethical reasoning skills include (a) the ability to determine the relevance of an issue to ethics, (b) the ability to weight competing values, and (c) the ability to identify mistakes in value assessments (Tuana, 2014). These skills are only possible through the application of reason and capacities of all and only moral agents.

This means that there are no simple and formulaic answers to moral problems in digital ethics or in any other moral realm. Ethics is much more complex than the dogmatist's stance that there is one and only one way in which to conceive of a state of affairs or the solution to a problem. This is not to say that "anything goes" in ethics, or that ethical theories are somehow interchangeably compatible. Rather, it is to say that there are ethical issues and cases that might be more adequately addressed and attempts made to solve it from a utilitarian point of view in one case, or from a contractarian view in another, or from some combination of theories in another. It is also to say that we must recognize that there are different stakeholders, including human and non-human animals and other things whose interests and concerns may need to be taken into account in our attempts to theorize and on which to act in ethics. The binding force of theories about reasoning ethically is the identification of shared values that act as starting places.

In some contexts, like bioethics, these starting places have been codified as *principles,* kind of mid-level norms on which reasoning can act. In short, actions and the principles, ideals, and moral orientations upon which actions depend or upon which they are based are not simple and formulaic in our "real-life" attempts to solve moral problems. Accordingly, the interests, concerns, knowledge, and moral convictions of those who have an interest in those problems are rightly taken into account to find ways to solve the problems that are acceptable to all those concerned, and that will in the best possible cases for this book to lead us to actions that conduce to the use of digital technology for social good. To implement this pluralistic approach is to recognize the applicability of any single theory or combination of theories in identifying, analyzing, and arguing about cases in digital ethics. The point is to try to solve practical problems in the digital realm, not simply to "win" an argument against someone or a group who does not agree with you. Because this is the case, it is important to take a collaborative approach to engaging with digital ethics, and to argue as carefully and fully as possible for positions that it is reasonable to take. Doing this is a matter of intellectual honesty and moral integrity, two virtues we hope you will experience and practice in your personal and professional life, not merely develop and hone in a course on digital ethics or from reading this book.

In the next section, some examples of the use of various theories, points of view, and the need for fact-finding are presented to demonstrate the value of the pluralistic approach we take and advocate in this book.

Applying Theories in a Pluralistic Approach

Practical solutions to problems are solved or considered from any number of different points of view and theoretical backdrops. This does not, as mentioned previously, mean that "anything goes." The most important elements of this manner of conceiving of practical, applied ethics are (1) a virtue ethics conception of our place in communities and the development of moral character; (2) the need for reasoned inquiry into the facts and facets of issues as they present themselves; (3) an investigation of the nature of the moral problems to be solved; (4) consideration of the human and nonhuman animals, environments, or things affected by our decisions and that their (and our) interests are properly taken into account; and (5) recognizing that if we will work together, agreement on solutions can be reached even among individuals whose theoretical loyalties are different from our own.

Real-life moral decisions are not always easy to make. There are contingencies, facts, disagreements, emotions, persons, places, things, and conditions to consider. Perhaps, in the best possible case, everyone involved agrees or every theory used to think about a case leads to agreement that a proposed solution or action is the right and good one. But real-life decision making is not likely to yield 100 percent agreement. Consensus is a more likely outcome where even the person who does not agree with the decision has had her or his position respectfully considered and taken into account. It is then that we can all *accept* the decision made even when someone does not *agree with* the decision. That is the goal of a pluralistic approach to ethical decision making and acting. It is not to create conditions in which the majority rules, or where "might makes right," but instead, it is to create conditions in which the best decision, even if not the most popular or ideal decision, can be reached.

The means to achieve the best decisions and actions is through discussion, using the interplay of ideas and opinions, respectful consideration of viewpoints, and facilitating this process through reasoned discourse using the best possible information we can gather. The deontologist may not be able to make a distinction between persons because every person is a being with inherent value, and the utilitarian might not be able to decide in an emergent case which person's life being saved will achieve the greatest happiness; the virtue theorist may not be able to determine how virtue of character will lead to a decision, and the adherent to the ethics of care may be at a loss in some situation regarding the interests of persons and the interrelatedness of individuals. But all these points of view may be given proper respect in making a decision by considering the questions they may ask and the concerns they express regarding the moral problem at hand. The deontologist will ask what is the right thing to do, regardless of consequences; the utilitarian will ask about the possible and probable results of the alternative decisions presented, and subsequently decide based on the benefits and liabilities of the decisions; the adherent to the ethics of care will ask, as will the virtue theorist, how different decisions will affect individuals and the communities of which they are a part. The pragmatist and the contractarian will seek consensus; they will seek the solution, in Dewey's phrase, that "will do." So, what is the solution to a moral problem? To find out, talk to the people who are involved, consider their opinions and positions, present your own, and weigh the alternatives with them. We cannot provide the "right" answer to any question or to any of the cases we present for your consideration in this book. Only you and others involved in moral decision making in particular cases can decide -- and you can do that only by communicating with them.

Suppose, for example, that you are responsible for deciding whether to sacrifice the well-being of one person for the well-being of another. Your dilemma is in trying to decide whether to save the life of one person with a single available dose of a nanotechnology medicine that cannot be split or shared, and because of the fact that it cannot be divided, another other person will not receive it. There is one dose—for one person. How do you decide what to do?

Perhaps the first thing to consider is that unless there is a dire emergency and you are completely alone as decision-maker with two people in need of the medicine, there will be others whose opinions and positions need to be considered. The viewpoints and interests of the patients themselves are surely relevant to consider, their relatives' concerns are a factor, and other people, perhaps hospital administrators, physicians, and insurance company representatives, have positions that deserve attention. How will all of you make the decision? And even if you are the lone decision-maker, should you simply look at the problem from, say, a deontological point of view? What about a utilitarian perspective? Perhaps the first question in tackling the problem is not who is to receive the medicine based on steadfast adherence to a specific theory, but instead how will you go about making the decision concerning who will receive it.

For the deontologist, you cannot make your decision on the basis of whether one of the people has a life more "worth living" than the other. You must base your decision on respect for persons. Both human beings are of equal worth, so your decision to whom to provide the medicine cannot be made by considering, for example, that one of them is a famous physician developing a cure for cancer and that the other is a college dropout with a criminal background. From a utilitarian point of view, which person's receipt of the medicine would be most beneficial, overall? Perhaps saving the life of the famous physician seems the most worthy use of the medicine, but there is no guarantee that she will ever develop a cure for cancer, and perhaps the dropout with the criminal record is also a wealthy philanthropist who regularly provides funding for cancer research—or his plan is to do so now that he won the state lottery last night. Working from any moral perspective, there are various considerations to evaluate. Is there some common element among the competing theories that will tip the balance of the decision in one direction rather than the other? The way to find out is to consider the situation in which you find yourself from many angles. The first angle is facts.

Wouldn't you need to know which of the patients has the best chance of surviving if given the medicine? Perhaps the famous physician is too weak and ill to be likely helped by administration of the medicine while the dropout-criminal-philanthropic lottery winner is not as sick, is young, and is almost certain to survive if given the medicine. Even if all the physician's friends and patients prefer that she be given the medicine on the off-chance that she will survive, the most practical course of action is to give the medicine to the budding philanthropist—even if he changes his mind later and decides not to be philanthropic at all. Considering the scarcity of the medicine matters as well, as do facts such as the likelihood that another dose of the medicine will be available later, or that alternative treatments for the physician can also be tried.

The point at this juncture is that the facts of a case may render the decision easier to make because the viewpoints from moral theories are overridden by practical considerations. In such a case, it is perhaps ironic to note that comparative moral theorizing has little to do with the decision to be made. It seems likely here that no matter any person's preference for a theory, the lottery winning criminal drop-out "wins" from all theoretical points of view. Perhaps more ironically, it is from the point of view of a theory (pluralistic, pragmatic ethics) that the facts have made all the difference.

The nature of a case may well determine whether it is facts that make the decision for us, or whether it is our theoretical loyalties or propensities that are the most prominent factor in making a decision on a moral issue. In digital ethics, one of the most intractable problems is lack of civility in online communities caused by the behaviors of people and the effects on others that online communities and social media can have. Regardless of one's adherence to her preferred ethical theory, one may wonder why many people are so willing, ready, and able to engage in ugly verbal disputes online when often in their personal lives outside the virtual world, they are nothing like that. Online communities have become central to the problem of online bullying, especially among vulnerable pre-teens and teenagers, sometimes leading to suicides and other forms of violence.

Whether you are a virtue ethicist, a contractarian, a deontologist, a utilitarian, or a care ethicist, you are likely concerned regarding what to do about problems inherent in online communities. As a virtue theorist, you may focus your attention on what we can do to improve the online lives of social media users. As a deontologist, you may wonder why people do not treat each other with respect. As a contractarian, you may consider the strength of user agreements and whether people will adhere to them, and what to do if they do not. As a care theorist, your primary concern may be to devise some means by which to handle the problem of incivility and its effects by creating stronger ties between members of the community.

Regardless of the theory you adopt, it seems reasonable to believe that we can all agree that something must be done to curb or eliminate online bullying. So, the issue in this case is not whether online bullying is problematic. Reasonable people will agree that it is so. On the other hand, what action should be taken, and which facts, considerations, and theories are to take center-stage in deciding what is to be done to control or stop online bullying?

To reiterate, our purpose here is not to tell you the solution to the problem, in this case bullying in online communities. It is instead to describe the process by which you and relevant others may go about trying to solve it. Interestingly, it may be at the outset that people disagree about the nature of the problem. Everyone might agree that online bullying must be stopped because it is in itself morally wrong from any theoretical point of view. But stakeholders in the case may disagree about whether bullying has even occurred in some specific case, whether there should be limitations and exclusions on the people (perhaps age-determined) who might be permitted in certain online communities, whether discussions should be moderated, and so on. It might not be, given such considerations, that bullying itself is the problem to be solved. Bullying may instead be a symptom of a problem such as lack of civility in discussions, lack of care for community members, failure to adhere to user agreements in an online community, or lack of appropriate respect for persons. Identifying the problem, in short, is the—or at least a—first consideration in evaluating and trying to solve a problem.

Assume that you and others have agreed that while bullying is surely problematic and it is the impetus behind your discussions to try to find a solution, you have all agreed or determined that there should be age-limits placed on membership in certain online communities depending on the topics of discussion and the purposes behind a community's existence. Suppose, for example, that the conclusion has been reached that children under the age of 18 ought not to be permitted access to a certain online community due to the nature of sensitive discussions taking place and that users must agree to end-user license agreement (EULA) terms that include a statement of standards for discussion and those who are to be permitted in the group.

What further thoughts need to be taken into account in implementing these two conditions for entrance into a discussion group or online social media group? It is impossible to give a comprehensive list here, in this book, because it requires the actual presence of people who are already involved in the group, their concerns, the general topic of discussion and other online activity, the facts they have presented regarding the issues, and their theoretical leanings. The goal is to reach consensus, and thereby formulate a plan of action guiding the users of the online group or social media platform that protects its users. But even then, issues may continue to arise based on the concerns, interests, and points of view of users and other stakeholders. Perhaps an additional concern is the notion of "protection" of users overall. There may be members of the community who find the notion that they will be "protected" from ideas or discussions to be antitheti-cal to their ability to make decisions for themselves (avoiding paternalism), and perhaps that "protections" in online communities lead to censorship, violating rights regarding freedom of expression. These are all issues and concerns that are evaluated and analyzed in discussion with others, considered from various points of view, and a solution reached and implemented. The point is ultimately one of gathering information, listening to legitimate concerns, applying various points of view and ethical theory-based opinions, and finding common ground on which to act.

Interestingly, it is often the case with respect to ethical theories that some members of communities are not content with consensus and working in con-cert with the theoretical or ideological views of others and wish their positions to be accepted by all. In this book, however, given the pluralistic, pragmatic approach that we advocate, such absolutist tendencies are antithetical to the solution to problems. Consider, for example, that for the virtue theorist, there are no principles or "rules" of action to be applied that will be satisfactory since the goal of the virtue theorist is to take into account the character of actors, and the tendency of one's actions that spring from character, that lead to the goal of creating and managing a community. While the virtue theorist may sincerely believe that everyone ought to adopt virtue ethics as a guide for life, the sim-ple fact is that not everyone is a virtue ethicist, and guidelines and rules and procedures for the management of an online community or social media group may be the best solution we can achieve given the many and varied points of view coming from utilitarians, deontologists, contractarians, and others. We must therefore be content, if we wish to solve problems and not to argue end-lessly with each other over theories and ultimately leave problems unresolved, to work with others, to respect and listen to their positions, and find points of consensus on which we may all act regardless of our most deeply held and cherished beliefs and theoretical loyalties. In essence, we must all be willing and able to set our own beliefs and theories next to those of others and find common ground because, after all, our concerns are the same: to identify and to solve moral problems for the good of individuals and for the communities of which they are a part.

Next Up

In the next chapter, we tackle the third element of moral literacy—that of moti-vation to act. Motivation is complex and difficult. We offer case studies as a means by which investigators into moral problems may develop habits of char-acter in a virtue-based, pluralistic approach to ethical inquiry. This encourages them not only to recognize moral problems, but as members of a community, to be emboldened to act on them.

References

Aristotle. (1999). *Nicomachean ethics* (2nd ed.). (T. Irwin, Trans.). Indianapolis, IN: Hackett.

Dewey, J. (1920). Reconstruction in moral conceptions. In *Reconstruction in philosophy* (pp. 161–186). New York: Henry Holt and Company.

Gilligan, C. (1982). *In a different voice: Psychological theory and women's development.* Cambridge, MA: Harvard University Press.

James, W. (1897). *The will to believe.* New York: Longmans, Green, & Co.

James, W. (1995 [1907]). Pragmatism's theory of truth. In *Pragmatism* (pp. 76–91). New York: Dover.

Kant, I. (1993). *Grounding for the metaphysics of morals: With on a supposed right to lie because of philanthropic concerns.* (3rd ed.). (J.W. Ellington, Trans.). Indianapolis, IN: Hackett.

Kohlberg, L. (1981). *Essays on moral development Vol. 1: The philosophy of moral development.* San Francisco, CA: Harper & Row.

Lipman, M. (1987). Ethical reasoning and the craft of moral practice. *Journal of Moral Education 16*(2), 139–147.

Locke, J. (1980). *Second treatise of government.* (C.B. Macpherson, Ed.). Indianapolis, IN: Hackett.

Midgley, M. (2000). Biotechnology and monstrosity: Why we should pay attention to the 'Yuk Factor'. *The Hastings Center Report 30*(5), 7–15.

Mill, J.S. (2001). *Utilitarianism.* (George Sher, Ed.). Indianapolis, IN: Hackett.

Peirce, C.S. (1877). The fixation of belief. *Popular Science Monthly, 12*, 1–15.

Rorty, R. (1989). *Contingency, irony, and solidarity.* Cambridge: Cambridge University Press.

Rossman, D. (2017, June 20). Texting suicide trial: Will guilty verdict in Michelle Carter case change laws for good?" *Newsweek.* Retrieved from https://www.newsweek .com/words-kill-teenagers-text-message-incited-her-boyfriends-suicide-should-she -be-627535.

Tuana, N. (2014). An ethical leadership developmental framework. In C.M. Branson & S.J. Gross (Eds.), *The handbook of ethical educational leadership* (pp. 153–175). New York: Taylor and Francis.

3

MOTIVATING ACTION IN DIGITAL ETHICS

In this chapter we discuss the problem of moral motivation as the third component of digital ethics literacy. We describe the relationship between reasoning and motivation, clearly define ethical motivation, and offer case studies to unpack some of its core concepts. We then argue that a virtue-based analysis of the psychological problem of moral motivation, taken as an important part of our pluralistic approach to ethical reasoning, best aligns with the interdisciplinary contexts of a diverse digital world. This strategy enables thinking about the relationship between individuals and their communities. Carefully designed case studies can provide effective modes by which to engage in conversation and motivate us to act on ethical decisions by cultivating habits of inquiry and developing moral hope, purpose, and courage by which to act.

Moral Motivation

So far, we have argued that recognizing digital ethical issues and their stakeholders are virtuous goals, yet these goals are incomplete without the ability to reason *thoughtfully* about ethical issues and the contexts in which they unfold. Further yet, reasoning about ethical issues is not sufficient for actually *acting* upon them. Ethical motivation involves building up the right sort of moral character (through the cultivation of virtues like hope and courage) but also overcoming psychological, social, and personal roadblocks to action.

CASE STUDY: VOICE DEVICES AND PRIVACY

A quick look at an Amazon account with which an Alexa device is associated will show a listing of every command given to the device including voice recordings from mis-heard commands that were not intended for the device. Examples include television commercials playing in the background in which the "wake word" or a similar spoken sound is noted by the device, and a short (and sometimes not-so-short) recording of the background television noises and the voices of people in the vicinity of the device capturing their conversations as a result of the device "listening" for the wake word. These recordings appear along with every command intentionally given to the machine, from the command to play a movie, television show, or music to a query regarding the time or temperature. One example noted in the news (Weise, 2018) is of a Portland, Oregon family's private conversation being recorded by an Echo speaker and emailing it to a person from the owner's contact list. The friend from the list contacted the owners to let them know that their private conversation had been sent to him.

Concerns continue to be raised about privacy, spying, and sensitive information being saved, stored, and sent without the knowledge or consent of the owner. Concerns about the use and capabilities of Amazon Echo and other, similar devices such as Google Home and even smart watches have raised concern in Germany, where smartwatches and Internet connected toys are being banned (Lomas, 2017) due to privacy and safety concerns for children who have them. Other concerns relate to the privacy of others (such as parents who are eavesdropping on teachers in schools through their children's connected devices) who may be recorded without their consent. In addition, an August 12, 2018 article (Greenberg, 2018) reports that there is a new security hole in Echo devices that will allow a hacker to "hijack" the device. Only in August of 2019 did Apple and Google agree to temporarily stop their employees and contractors from reviewing recordings from their home automation devices (Tung, 2019).

Perhaps a good question to ask, given complex cases like in-home digital assistants, is how reasonable it is for users to expect developers to be required to be aware of every "side-effect" or unintentional effect of the use of their devices. Another reasonable question in the Echo case is whether it is possible to close all the "holes" in programming and render the devices completely secure from hacking or from undesired dissemination of personal information. Asking such questions about the case will lead you, if you are careful and diligent, to seek out articles and books on the issues and questions discovered, resulting in the ideal case of the investigators being more fully informed and therefore more fully capable of offering reasonable solutions to the moral problems involved. If, for example, an investigator wrongly believes that it is possible for a software developer to create software that is completely immune to hacking or to undesirable uses, it will be impossible to offer solutions to the problems of Internet-connected devices that are practically possible, short of making the obviously questionable and unjustifiable pronouncement that it is not morally acceptable for anyone to use such devices at all.

But you can imagine the complexity involved in this sort of careful research, especially given the speed and scale of digital information technologies. The attention and energy required in cases of digital ethics can not only lead to inaction, but to *amotivation*, a feeling of helplessness in the face of ethical concerns.

Reasoning to Action

So what is the relationship between ethical reasoning and moral motivation? One of the central goals of this book is to make it clear that moral decision-making and acting on such decisions are not separable activities. To make a moral claim, judgment, or decision is properly and completely understood to prescribe action to accompany it. While it is certainly possible and perhaps common for people to make judgments about their own actions or the actions of others and yet intend to do nothing, there is much more to ethics than talking and theorizing. Action is also required.

As we described in Chapter 2, ethical reasoning in the digital is wickedly complicated. Digital ethical reasoning faces what we might think of as a processing power problem. If we are right in arguing in previous chapters that one of the (if not *the*) unique attributes of the digital is radical change in speed and scope, then the information flows that inform ethical decision making move faster and more broadly in the digital than in analog forms. Indeed, they now move at greater scale than the human mind can process. The human brain (if we think of it as the analog to a computer) processes with billions of neurons working together in parallel, so it is capable of processing *speeds* far greater than is possible for any other computing system. Similarly, the *scale* of information available to process is overwhelming: doubling every two years (de Planque, 2017). This radical shift in scope denies or limits human opportunities to process information, form and evaluate knowledge claims, and assess value conflicts in a short enough time to reach decisions about problems before their conditions have shifted. Simply put, we are more easily overwhelmed by the sheer amount of information passing by, around, and through us. To offer a simple analogy, imagine the way we processed news in analog form in the pre-Internet era. Most of us had one or two national news channels on television, one or two major newspapers, and our local communities with whom to exchange ideas, learn of new and emerging problems, and form judgements. These "limitations" simply allowed us to *perceive* information at a more reasonable pace, giving us the mental space to cognitively process information, form opinions, and make judgements. Now we have access to dozens of news outlets, each with a slightly different "take" on the same stories, alongside endless digital publications and boundless global digital communities. The breadth, speed, and specificity of context to which we each have access is now substantially and radically different in the digital. At some point, our capacity to process information (along the way to making ethics decisions in reasoned ways) is overwhelmed by the amount of information available to us.

Given this processing power problem, deciding that what we are about to do or what we have done is the right or good thing to do or to have done is a difficult process, especially given that there are variations in the psychological motivations we have. It may be that our motivations are grounded in the theoretical ethical points of view we adopt. Often, it may also be that one is motivated to act outside such specific boundaries of theory and instead we are motivated to act out of anger, love, or some other emotion. There is no last word on which of these sources of motivation is justifiable, and we discussed many of the theoretical complexities that may inform motivation in the last chapter. But one thing we can say with some certainty is that these two sources (and perhaps there are more) of motivation can be categorized in the first case as rational motivation and in the second case as non-rational motivation.

We think that both motivators are morally justifiable but that they are not justifiable in and of themselves. That is, even if our motivation to act is based on the principles of some ethical theory, it is not for that reason alone to be considered justified. Instead, our reasoning within a theoretical framework requires argumentation. The same can be said of a non-rational source of motivation (an emotion), where it is necessary for us to justify the actions performed based on emotion to ensure that the actions are themselves reasonable.

It is surely reasonable to have emotions and to have emotional reactions or feelings regarding an ethical issue or problem. Going back to the "yuk factor," for example, we might feel the emotion of disgust and revulsion at a certain course of action, but as we have already seen, the simple fact that someone feels repulsed by an action or occurrence does not, in itself, mean that it was the wrong thing to do or the wrong thing to have occurred. Simply put, we have to argue for the positions we take and the actions we perform based on those positions. It is therefore the case that reasoning well about issues is essential to understanding the nature of a moral problem and is necessary to propose and to act on any proposed solutions. We take it as basic that the only reasonable approach to identifying and attempting to resolve moral problems is to argue for positions.

Our pluralistic view of practical ethics (developed in Chapter 2) is in line with the interdisciplinary and diverse landscape of the digital. Pragmatic thinkers use normative theories as tools for decision making, or as guideposts by which to stay true to a general course of thinking. Questions of consequence matter, but balance against concerns about rights and duties; relationships inform individual choice, and matters of virtue become broader questions of integrity. This is just as messy as it sounds; but, of course, the ethical issues we face are themselves messy—messier sometimes in the digital, we think, than in analog contexts.

In analog contexts too, ethical reasoning has its share of problems. Just try to imagine a world in which we all fully exercised our capacities to reason all of the time. In such a world, ethical problems may arise but would swiftly be acknowledged, understood, and analyzed toward actionable consequences. We do not live in a world like that. In fact, sometimes it seems to us like humanity has for-gotten its rational capacity and acts out of caprice, pride, raw emotion, and mere sentiment. Some philosophers have thought similarly, well before us. British empiricist David Hume, for example, thought this was right: He thought senti-mentality, not reason, was the driving force of our decisions (Hume, 1739). Even further back, the ancient Greeks knew this to be a possibility, arguing that the human soul was split between emotion, reason, and a middle ground of caprice, pride, and raw response they called *thumos* (which one of Beever's undergraduate professors once described as that feeling you get when you stub your toe on a chair and get so angry you want to kick it with that very same foot). And here in the early 21st century, the focus on affect (emotion and sentiment) has returned in the work of the moral psychologists, who use social science techniques to evi-dence ways in which emotions and *not* reason guide human action (e.g., Haidt, 2013). So, one problem with ethical reasoning is that it is not sufficient for ethical decision making but *merely* necessary. Neither the Greeks (generally speaking), nor Hume, nor Haidt would likely disagree with our claim that reasoning plays an important role in making decisions—their shared point is just that it *alone* isn't important: affect matters too. Reasoning requires affective virtues like courage, hope, and purpose to move from in the head and the heart out into the world. Nancy Tuana describes motivation of this kind as a sort of *willing*. "If education leads to the *ability* to read," she writes, "but not to the *will* to read, then educa-tion has failed" (Tuana, 2014, p. 169). Action in the world is the goal of moral motivation.

The virtues of courage, hope, and purpose drive ethical motivation. Moral purpose is the wellhead from which ethical motivation springs: It is a com-mitment to a clearly reasoned set of ethical values and goals, either at the indi-vidual level or as shared by a community. The purpose from which we act sets the direction of our action. Some ethical actions require directed moral cour-age—like saving a drowning child if it would put you yourself at risk of drown-ing—but simply "living by our values take tremendous courage" (Tuana, 2014, p. 171). Having *integrity*, of the right kind and in the right way, is just to live by your (morally righteous) values despite hardships they might entail. There is nothing easy about this, and the results can be frightening. Courage is required. Similarly, we cannot be ethically motivated without hope. Without hope, we face amotivation, driven by despair and cynicism. For example, we could give up the fight against anthropogenic climate change since scientists agree we are committed to some levels of radical ecological change already. Or we could hope for a biodiverse future, and be motivated to act by that hope.

Acting on moral problems also includes proactively working toward personal and social moral good. To tell one's friends and children that they ought to be honest in their online dealings with others is more than simply to utter a state-ment about how we ought to act. It is also, in the most complete sense of living a moral life as a member of a community, to prescribe that such behavior is good for ourselves as well as others. It is a morally bankrupt person and society who recognizes a moral problem and fails to lift a finger or to say a word to try to solve it.

Imagine, for example, knowing that your neighbor violently and viciously beats his children but you have decided that while you think it is morally unconscionable for such behavior to take place, it is none of your business and none of your concern. As a result, you neither call the police nor offer assistance to the family in any way and instead close the doors and windows of your house to make it less likely you will see or hear any activity from the neighbor's home. Behaving in this way may be comforting to you because you have closed yourself off from the problem taking place next door, but it is certainly not conducive to solving a serious problem experienced by other human beings, and it is a sad testament to the state of your character.

This hypothetical example gained notoriety in real life in the case of Kitty Genovese who, in 1964, was attacked and murdered outside of her apartment building in Queens, NY. The *New York Times* later reported that 38 individual people heard her cries for help—and did nothing. Each claimed that they assumed someone else was going to intervene, call the police, shout out the window, or drop a grand piano on the culprit's head. Psychologists went on to name this result the "bystander effect," despite later evidence that the case was incorrectly reported (Manning, Levine, & Collins, 2007). Now, imagine a similar case taking place in 2019, where every one of those observers had a smart phone close to hand. Instead of dialing a home phone, they could dial the police with a single button while simultaneously streaming video to social media feeds. Would this be a deterrent to the crime? Would it have alleviated the "bystander effect"? We have no easy answers to questions like these, but the differences between analog and digital events is a key consideration of digital ethics.

Whether analog or digital, that failure to act in such a case of moral tragedy is ethically bad—and this badness aligns with our common moral intuitions. The implications of this position have been articulated in compelling ways by philosopher Peter Singer in his well-known 1972 article "Famine, Affluence, and Morality." There, Singer argues that we have a duty to rescue in every case where we are not required to sacrifice, in his words, "anything of comparable moral importance" (Singer, 1972, p. 231). He offers the example of the drowning child to explain: "If I am walking past a shallow pond and see a child drowning in it, I ought to wade in and pull the child out. This will mean getting my clothes muddy, but this is insignificant, while the death of the child would presumably be a very bad thing" (p. 231). That seems eminently reasonable, and it sets Singer up to argue that it makes no morally relevant difference if that child is close by or across the world—our duty to rescue holds in all cases. Our digital technologies give us access to children across the world in ways (good and, unfortunately, bad too) that Singer could not have foreseen—and so challenge claims like his about the scope of moral responsibility.

With all of this said, there are sometimes contingencies to action that arise based on times, cases, places, and circumstances that impinge on the way or ways in which one may go about acting on ethical issues and ethical decisions. If you have been threatened by a neighbor who beats his children, you may reasonably hesitate to go to the neighbor's door to confront him about his behavior, but that should not stop you from calling the police. On the other hand, if your neighbor is your brother and you know that his behavior is the result of a medical condition, you may determine that the proper course of action is to call the police, but not with the goal to have him arrested. You want to assist your brother and his children by helping to ensure that he receives proper treatment for his condition and that the children are placed in a safe location until your brother has returned to normal.

There is no one ethical theory capable of telling you exactly, in every case, what you ought to do to solve a problem or to avoid one. Nor should we expect a theory to do so. Instead, theories, principles, and concepts in ethics are but one element in the process of analyzing, evaluating, and acting on moral decisions. In adopting a pluralistic approach to ethics, we take the position that it is largely the responsibility of the individual to determine the right thing or the good thing to do, and to act appropriately based on one's best judgment when a situation calls for individual action. But even here, the individual's best judgment is informed by current or previous discussions and decisions of a community taking into account the vagaries of experience in any particular cases. The best judgments we are capable of making are the ones that are informed by appropriate facts, based on consistent and applicable concepts, are the result of careful reasoning, and that take into account the effects of one's actions on other individuals, the community, and the society in which one lives.

Contingencies aside, sometimes moral motivation faces external challenges of a different kind. These describe the problem of *amotivation*. Under what conditions do we know that helping the drowning child is our moral duty, yet we fail to act on that duty? Amotivation has been defined simply as "the state of lacking an intention to act" (Ryan & Deci, 2000, p. 61). Intention to act, or motivation, arises from an alignment between one's capacity to reason and one's capacity to act on those reasons: between knowing what the right thing to do is, and doing it. Of course, *amotivation* is a problem because of the wide range of external factors that limit our actions, or will to act. Imagine, for example, a corporate whistle-blower who realizes that their company is polluting a nearby waterway by unethically dumping electronic waste they produce, threatening not only ecosystemic health but also human health downstream. We can assume that this person knows the right thing to do is to "blow the whistle"—to call out the company in order to get it to change its actions. They know e-waste is a substantial problem at the intersection of digital tools and analog environments. Yet this hypothetical person—and real people in similar situations, too—often find whistle-blowing an extremely challenging act.

They might worry about repercussions, job loss, industry blacklisting, harms to family and friends, feelings of betrayal, et cetera. All of these concerns threaten to override their own rationalization that bringing forward evidence of the pollution is the ethically right thing to do. Those individuals who are able to overcome those external constraints do so with *moral courage*. Indeed, in her model of moral literacy on which we lean in this book, Tuana (2014) articulates that moral motivation consists of at least three interwoven components: moral courage, moral hope, and moral purpose (p. 169). Together, these virtues, or characteristics of the individual, facilitate the link between ethical reasoning and moral motivation.

So, how are these concerns about moral motivation changed or challenged in the digital? In our view, the challenges to ethical action are much the same, but more complicated and quicker moving in the digital. In some ways, the implications of our actions are wider-spread through the accelerating flows of information in social media and networked streams. In other ways, the implications of our actions are potentially lessened, surrounded as we are by a nearly boundless diversity of views and opinions, all clamoring for validation and screen real estate. Imagine the same whistle-blower from our earlier example. Under analog conditions, their story might end up contained between a regulatory body and the corporation, and it might make a newspaper inside-fold story after a few months. Implications for the corporation, and its shareholders, might take months to begin to emerge, and could be countered through measured marketing strategies. Under digital conditions, however, whistle-blowers can share stories with global communities and regulatory bodies instantaneously. Stock prices, now governed more by algorithm than by human trading, may respond immediately to the first stories of possible unethical corporate action. The name of the game in the digital, one might worry, is that motivation is becoming less the result of careful reasoning and more the result of manipulation of affect and designed context. Take, for example, the September 2018 Nike campaign featuring Colin Kaepernick, a name only on the national tongue thanks to his "take the knee" campaign against police violence on minorities. Nike had to have known of the divisiveness of Kaepernick's name and image, yet they chose him as the face of a major campaign anyway. Was the purpose to drive shoe sales? To take a stand against police violence? To lean a little to the political left in the hope of drawing in new customers? Any or all of these might be the case because the gamble of Nike was the digital version of the old saying, "any news is good news." Manipulation of digital information streams is what motivates—it is what draws attention and focuses thought. Whether our position here has any ethical weight is something to be determined on your own. But we think it is a clear example of the ways in which moral motivation, like ethical decision making writ large, is increasingly complicated in the digital.

To complicate matters further, we do not only face moral problems of our own, but we are also involved in the moral problems of other people. As spectators rather than moral actors, we often make judgments about the attitudes and actions of others. For example, consider the following scenario:

CASE STUDY: MAJOR DATA BREACH

Suppose that a large corporation in the U.S. suffers a major data breach involving thousands of current and former employees. The data breach is announced to the public and to affected people about a month after the breach is discovered. Among the personal and financial information stolen are full names, employee ID numbers, social security numbers, home addresses, and W-2 forms. The corporation provides to those affected one year of identity theft protection at no charge, assurance that law enforcement is involved in an investigation into the breach, and a promise that protection of personal information is of the utmost importance to the corporation. From your point of view as a new victim of identity theft, however, you think that a year of credit monitoring or identity theft protection, a vague claim that law enforcement is looking into it, and that the company cares about your personal information are insufficient responses. From your point of view, what are the moral problems involved in the data breach and in the corporation's response to it? How would you go about evaluating the problems you have identified? Who are the stakeholders in this case, and what should they do? What moral point of view—or moral points of view—are you using to identify, to evaluate, and to propose potential solutions to the problems you have identified?

It is likely, in the case above, that the corporation's representatives who are responsible for responding to the data breach have a view of the matter different from yours and others whose personal and financial data were stolen. Those outside the corporation, such as the general public that was informed in newspapers and on local television news, may have yet a different view. For all those involved, there are issues of responsibility and considerations of the effect of the data breach on the community and the duty of the corporation and its officials to handle the data breach in a fair and equitable manner. Additional questions regard the potential and actual consequences of the data breach and reactions to it concerning the people involved. Ultimately, the most justifiable moral course of action here is to carefully evaluate the various effects of the breach in order to determine what level or type of consideration is appropriate to extend to those adversely affected. Moral actors must be motivated to respond to occurrences such as data breaches, and this requires the development of character and a social structure consistent with open communication and respect for ideas.

As we previously noted, it is one thing to be aware that there are moral issues in digital technology—whether in its development or in its use—and it is a completely different thing to wish to do something about those issues. Further, even the ability to recognize moral issues associated with digital technology is itself fraught with problems since, for example, the desires of the user or creator of digital technology may be so strong that the user or developer fails to take notice of unintended consequences or uses of the technology. It is likely not the case that the developers of tablets and cell phones and other connected devices designed them to be used to sell and distribute child pornography or to commit financial crimes, but the simple fact is that those devices are used for nefarious— as well as for good—purposes. Making the distinction between the good and the bad is sometimes difficult, and being able to recognize in some less clear or obvious cases problematic areas of their use is even more complicated.

A Virtue Ethics Approach to the Problem of Motivation

But how do we come to perform actions or to have appropriate motivation to act on the use of digital technology for social good? How do we extract our-selves from the temptation to create, say, a videogame with repulsive and mor-ally objectionable material when the potential for individual or corporate profit is high? How do we even know what is the social good? How are we motivated *appropriately toward ethical ends?*

We propose that deriving answers to these questions is possible by adopting a pluralistic approach to moral action and reasoning at the forefront of evaluating, creating, and acting on arguments while adopting a *virtue ethics standpoint on the psychological problem of moral motivation.* The primary reason to adopt a virtue ethics approach is to consider the difficulties of determining what constitutes "social good." The reason to use a pluralistic approach to moral action is that it allows us the most robust and diverse possibilities to do the right and good thing.

Virtue ethics and pragmatic ethics, unlike normative theoretical approaches like utilitarianism, deontology, or contractarianism, are focused on and central to the concept of community or society. Utilitarianism, deontology, and con-tractarianism are steeped in an individualistic tradition in which each individual reasons independently about issues that may be a part of social good or social duties, but it is not necessary that the individual act for that purpose. The indi-vidual, instead, may focus narrowly on a subset of the population rather than on social ends themselves. Utilitarian ethics focuses on individual goods which might be related to social goods but not necessarily so. Deontological ethics does not focus on the good at all, but instead its primary considerations are those of the individual's duty apart from consequences. And while contractarianism may broadly help us to conceptualize the social good on the basis of agreements we have made to constitute society in a specific way, its character is to center atten-tion on the individual human being and her or his rights. In other words, such theories are not necessarily amenable to considering fully or adequately the ways in which one's decisions and actions will affect the good of and for a society or community.

But traditional virtue-based ethics, the ethics of care, and pragmatism are oriented differently: They center on moral considerations derived from the society and community in which one lives. Aristotle put it well when he asserted in his *Politics* (350 B.C.E/1979) that the state is "prior" to the individual not in the order of nature, but in the order of becoming. While that statement overflows with metaphysical implications, it is an important one in recognizing human beings as primarily social or political animals, as Aristotle would have it, who are not all they can be if they live in isolation and contrary to their nature. We are not contending that there is some essential core of characteristics or behaviors that constitute "human nature," but we do want to argue that virtue ethics in its attention centered on community is more consistent with a pluralistic approach to the solution to moral problems than individualistic theories. A pluralistic ethics of virtue, care, and pragmatic considerations offers a rich perspective that can capture who we are as human beings. As rational beings living in communities, our attentions in moral and social matters must take proper consideration of the communities in which we live, and not simply to focus on the interests or goals of isolated individuals.

As social animals, human beings need our communities to realize goals. The community, in a similar fashion, needs the individual to exercise her or his virtues in making the community a better place. Alternately, Aristotle's position is that virtues are excellences of human character. The virtues are part of the means to living a good (happy) life. The virtues, however, are themselves socially situated and are perfected by practice. For example, a virtue of character is an excellence of a human being similar to the way that skill in ice skating is an excellence in coordination, balance, and graceful movement. For the most part, to become an excellent skater, a person has to practice, and practice requires a trainer, regular activity in pursuit of excellence, rules and techniques, the proper equipment, and judges who are experts who evaluate the performances of skaters. Excellence in skating, therefore, requires a community. The same may be said of the virtues as excellences of human character. For example, one does not develop the virtue of character called "honesty" by thinking thoughts about honesty or understanding the definition of the word. Instead, one becomes honest by performing honest actions. Even though honesty exhibited by or in one person may be different from honesty in another due to the conditions in which they find themselves, the other people involved, and so on, honesty is developed and perfected in the person over time and in conjunction with a community. For example, the honesty expected between spouses is different from honesty in conversation with a person you have just met. If last week you lost your entire paycheck gambling, providing that information to your spouse might very well be expected; but it is certainly not the business of a complete stranger what you did with your paycheck. Telling the stranger (if you tell the stranger anything) that you "spent it" is honest enough for the situation; telling your spouse (who is surely more privy to such information than the stranger) that you lost it gambling is perhaps appropriate in that context. The honesty you exhibited talking to a stranger in one case and talking to your spouse in the other about how you spent your paycheck are different. But in both cases, you are acting honestly, and both actions, even though they are different, are appropriate to the situation, the social context, and the relationship that exists between people.

When the habit of character that is honesty becomes part of a person—that is, part of a person's character—that person is said to be honest. Yet it is equally important to realize that it is not just the trait of character and the outward actions of a person that constitute honesty. It is also necessary that the behavior and the trait of character fit the conditions in which the person finds herself and that the action performed is not excessive or deficient.

We recognize, for example, that generosity in giving to charity for a person who is poor may be considerably different from generosity in giving exhibited by a multi-billionaire. For the person who earns very little, giving $10 to a charity is generous while for the person who earns millions of dollars a year, giving $10, while still charitable, is not generous. For Aristotle, there is a mean between excess and deficiency, which is a mean between vices, that constitutes virtue. Generosity is not too little and it is not too much. It is, instead, the exercise of the appropriate trait of character in giving to others the right amount under the right conditions for the right reasons, and so on. Generosity is therefore not the same action and it is not the same quantity for each person. It is relative to the individual, her circumstances, time, place, case, and abilities. The same considerations apply to all the human virtues, whether they are patience, courage, generosity, helpfulness, honesty, diligence, or those specified by ethical motivation (purpose, courage, and hope).

Further, friendship is essential to Aristotle's ethics and to the creation and maintenance of a good life and a good society. For our purposes, the various types of individual friendships that Aristotle enumerates and explains are less important than the more expansive experience of civic friendship that occurs in communities in which individuals share common interests, goals, experiences, and culture. Even so, understanding Aristotle's conception of friendship is useful to understanding the conception of civic friendship that helps to strengthen communities. Friendships of pleasure are short-lived and for a very specific purpose such as buddies with whom one plays cards on the weekends. Friendships of utility are also short-lived and usually end when a common goal has been achieved. People associating with each other to push a disabled car off the roadway or hunt down a Wi-Fi password in a public meeting place might experience a brief friendship tied to achieving that goal. But neither a friendship of utility nor a friendship of pleasure is the ideal of personal friendships. It is surely the case that a more lasting bond can be created between people from pleasure or utility, but it is a kind of friendship of mutuality that is the ideal in virtue ethics. Friendships of mutuality are the ideal in human relationships because they manifest themselves in feeling that the other person is akin to "another self" and that the friends are incomplete in some ways without each other. This is the sense in which private individuals recognize another person as a good friend or best friend, also recognizing their common interests and goals, and seeing each other as essential to living a good life.

Even though it is impossible for all people in a social setting to have the sort of feeling of mutuality experienced by "best friends," there is another sense in which one's community is best characterized by recognizing each other in the community as "civic friends" engaging in activities of civic friendship where their lives in a social group are intertwined and the contributions of each member ideally make the group better than it would be without the individuals. In a reciprocal fashion, the individuals are better because of the existence of the community because the community provides the individuals with the tools, experiences, education, and social structures in which the individual can develop her or his capacities (excellences, virtues) with others. Living in a community and recognizing the community in some sense as more complete than the individual does not negate the importance, the rights, the desires, and the goals of the individual nor does it make the individual subservient to the group. Instead, there is a mutually beneficial relationship between the individual and the community in which the community and the individual both progress and are made more perfect by their association with each other. The individual does not lose her individuality in a community. Instead, the individual experiences more opportunities for attaining personal happiness and the development of the virtues specific to herself in a community.

Conceiving of oneself as a part of the community of digital tool developers, users, programmers, engineers, and others who create and use digital technology, one has at his disposal the experiences, ideas, expertise, skills, and knowledge specific to digital technology that can and should create a bond between all. That bond, which is created and solidified through understanding the nuances and specifics of the development and use of digital technology, is like the bond between, say, physicians or professors or firefighters who work with each other to reach common goals within their professions and to solve common problems. The ideal case is realized when their knowledge, experiences, beliefs, and skills come together in dialogue among themselves and with stakeholders to create means to solve (and to avoid) the problems associated with their professions or placement in a group. It might be said that for creating and solving moral problems in all realms, including digital technology, many heads are better than one.

Avoiding Amotivation

Being a member of a digital technology community in being a creator, a user, a developer, a programmer, or having any other function in the community means that the role of each person and the experience and expertise she possesses is necessary to the proper functioning of the group. Different experiences and personalities inevitably bring disagreements, and rather than to see disagreements as something to be avoided, differing opinions and areas of expertise are beneficial to moral problem solving. In fact, they are integral to it. It is possible that one person may have "the answer" to a moral problem generated by the development and use of digital technology. Another person may have "the answer," too—but it is a different answer. When people are able to meet with each other to discuss the issues and work together toward solutions to problems, consensus may be reached and the interests and concerns of all are taken into account.

By recognizing the social nature of ethics and the social nature of the development and use of digital technologies, we may overcome roadblocks to action on perceived problems. When, for example, a team works together to develop a digital technology tool or program, respect for the voices of all involved, manifesting itself in everyone having the opportunity to express ideas to others, is a matter of developing the technology as well as ironing out potential problems before and after they occur. The same may be said of the ethical problems that appear sometimes as the result of the use of digital technology. If a member of a team recognizes a bug in a program, but that team member's concerns are not heard, the entire team suffers some setbacks as a result. The individual is therefore central to the workings and use of digital technology and we ignore the voices and viewpoints of individuals at everyone's peril.

There are numerous examples and cases of people in organizations who have reported problems in research and development and were ignored, only to leave an entire organization to experience fines, closure, or other sanctions from within an organization, from the public, or from government agencies. Whistle-blowers, for example, have traditionally been considered "snitches" and are shunned by others in an organization for uncovering and publicizing serious problems occurring within an organization. This occurs in contexts of secrecy and hierarchy in which one or a few persons noticing and reporting a serious flaw or problem in the organization is unwelcome. In a group of likeminded others whose goals are to create the best product they are capable of creating, and in a community of individuals who recognize the expertise and sincerity of each other, roadblocks to action after recognizing and reporting problems should not exist. Instead, individuals understand that their concerted efforts to solve a problem make both the product and their community better than it would be otherwise.

CASE STUDY: ACTIVE SHOOTER

Consider, as one final case study in this chapter, the development of the videogame, *Active Shooter*, which was modeled on school shootings. In a *Washington Post* article (Horton, 2018), *Active Shooter* is described primarily as socially and morally unacceptable due to its generally gratuitous violence and specific violence against women in the context of school shootings. Valve Software removed the game from its popular Steam platform (Steam is a gaming platform on which videogames like *Call of Duty* and others are available and through which the games are played), citing the unacceptability of the premise and development of the game.

Some parents of the Parkland, Florida school shooting victims expressed disdain about the game, with one parent pointing out that his daughter is dead in the real world and that this game "may be one of the worst." In the same article, a chief executive of an urban school coalition is reported to have written on Twitter that the game "violates every sense of decency a civilized nation should hold dear." Public outcry, in conjunction with corporate decisions on releasing the game, seemed to have effectively stopped its distribution through Steam.

In a *New York Times* article (Lucero II, 2018), Doug Lombardi, a spokesman for Steam, said that the decision not to release the game on the Steam platform was made by a group of people (although the NYT reported that there were no specifics given on who constituted the group) and that Steam would release "anything" except digital content that was specifically a troll and designed to cause controversy. Interestingly, then, it is not the content itself that seemed to be objectionable to the Steam leadership. Nonetheless, the content was and is objectionable to enough private citizens to generate opposition to its release, notwithstanding Steam's different concerns about trolling and causing controversy. Ultimately, the game is still available since the developer has provided free access to it. An article in *Fortune* (Morris, 2018) reports the developer, Anton Makarevskiy, expressing surprise that the game has received such negative reception and that people should focus on "the real issues" of school violence rather than on videogames.

In the case of *Active Shooter*, and in prior similar games like *Super Columbine Massacre RPG* (2005), the game developers seemed to have no moral qualms about the creation and distribution of their games, but members of the community of gamers and other interested stakeholders found their premises and content morally objectionable. But is it morally objectionable to develop such a game? How do we know that it is? What mechanisms are in place in our reasoning that lead us to the conclusion that the game is unfit for public use and distribution? How do we decide what to do about other games with violent content? And what about important concepts such as freedom of expression and the First Amendment? How do we balance freedom of expression and access to digital content with concerns about safety and public decency?

These, and many others, are questions that inform the cases we encounter in digital ethics. Solving the problems is not easy and levels of detail and sources of information and ideas complicate matters even more. Just what, we might ask, is an acceptable level of violence in videogames? And for whom? Our position is that it is up to the members of the community of gamers and others with an interest in gaming to discuss freely with each other their concerns and to provide reasoned arguments in an open forum of equals. This allows that community to act on principles and consensus to lead to the best possible decisions and actions that it is capable of producing.

The complexity of questions in cases in digital ethics demand careful reasoning. But while ethical reasoning is necessary for ethical literacy, it is not sufficient. While we might work academically through complex ethical issues and reason to important contextual agreement on the ethically right course of action, *performing* that right course of action remains contingent on being morally motivated. The virtues of purpose, hope, and courage orient our ethical motivation. And being motivated in this way is no straightforward task. Yet, as we argued above in this chapter, understanding the community-based nature of ethics is a first step in working through these potential problems in digital ethics.

Next Up

In the next chapter, we will begin Part 2 of the book. We start our investigation of the nature of digital ethics by exploring the properties of the digital considered through ethical perspectives. We will consider the distributedness of the digital as a keystone property, a conversation that considers the ethical implications of the speed, scope, and scale of digital information. We will also explore two properties of distributedness—reproducibility and transferability—and discuss how these characteristics present new challenges relating to agency, privacy, and ownership.

References

Aristotle. (350 BCE/1979). *The politics of Aristotle.* (E. Barker Trans and Ed.). London: Oxford University Press.

Greenberg, A. (2018). Hackers found a (not-so-easy) way to make Amazon Echo a spy bug. *Wired.* Retrieved from https://www.wired.com/story/hackers-turn-amazon-echo-into-spy-bug/.

Haidt, J. (2013). Moral psychology for the twenty-first century. *Journal of Moral Education, 42*(3), 281–297.

Horton, A. (2018). The 'active shooter' video game horrified Parkland parents. It was pulled before release. *The Washington Post.* Retrieved from https://www.washingtonpost.com/news/education/wp/2018/05/29/what-parkland-parents-think-of-a-new-video-game-that-lets-people-shoot-up-a-school/?noredirect=on&utm_term=.919de2cfb547.

Hume, D. (1739). *A treatise of human nature.* Retrieved from https://www.gutenberg.org/files/4705/4705-h/4705-h.htm#link2H_4_0085.

Lomas, N. (2017). Germany bans kids' smartwatches that can be used for eavesdropping. *Techcrunch.* Retrieved from https://techcrunch.com/2017/11/20/germany-bans-kids-smartwatches-that-can-be-used-for-eavesdropping/.

Lucero II, L. (2018). Steam, after pulling school shooter game, says it will sell nearly everything. *The New York Times.* Retrieved from https://www.nytimes.com/2018/06/08/technology/steam-games-active-shooter.html.

Manning, R., Levine, M., & Collins, A. (2007). The Kitty Genovese murder and the social psychology of helping: The parable of the 38 witnesses. *American Psychologist, 62*(6), 555–562.

Morris, C. (2018). 'Active Shooter' video game may still see the light of day, developer says. *Fortune.* Retrieved from http://fortune.com/2018/05/31/active-shooter-video-game-may-still-be-released/.

de Planque, T. (2017). Big data: Computer vs. human brain. *Stanford MS&E 238 Blog.* Retrieved from https://mse238blog.stanford.edu/2017/07/teun/big-data-computer-vs-human-brain/.

Ryan, R.M. & Deci, E. (2000). Intrinsic and extrinsic motivations: Classic definitions and new directions. *Contemporary Educational Psychology 25*, 54–67.

Singer, P. (1972). Family, affluence, and morality. *Philosophy and Public Affairs, 1*(1), 229–243.

Tuana, N. (2014). An ethical leadership developmental framework. In Branson, C. & S. Gross (Eds.), *The handbook of ethical educational leadership* (pp. 153–175). New York: Taylor and Francis.

Tung, L. (2019). "Apple, Google: We've stopped listening to your private Siri, assistant char, for now." *ZDnet.com.* Retrieved from https://www.zdnet.com/article/apple-google-weve-stopped-listening-to-your-private-siri-assistant-chat-for-now/.

Weise, E. (May 24, 2018). Alexa creepily recorded a family's private conversations, sent them to business associate. *USA Today.* Retrieved from https://www.usatoday.com/story/tech/talkingtech/2018/05/24/amazon-alexa-creepily-recorded-sent-out-familys-conversations/642852002/.

PART 2

The Nature of Digital Ethics

4

SPEED AND SCOPE OF DIGITAL INFORMATION (DISTRIBUTEDNESS)

Theorizing digital ethics should enable us to more carefully examine applied questions—or ethical issues that have implications in lived experiences in the world. In this chapter, we argue that *distributedness* marks a fundamental and novel property of the digital, one which brings along with it importantly novel ethical considerations for digital ethics. Distribution is a function of two attributes of digital information: reproducibility and transferability. These attributes stem from the binary codes that enable exact and unlimited one-to-one reproductions of digital documents and media. The speed and scope with which digital content can be reproduced and transferred is due to what James Moor called its "greased" properties (1997, p. 27). Such properties have profound implications, ranging from ethical questions about replication to deep ontological questions about authenticity, lack of scarcity, and modes of representation. These questions can be contemplated at varying levels of abstraction since they are relevant to individuals, societies, and professions. We will start with a playful case study to get us thinking about some implications of the distributedness of the digital.

CASE STUDY: HANDLEBAR HENRY

Henry, a young man in his late twenties, sports a man bun and handlebar mustache. He thinks that there is nothing like the sound of a cassette tape played on a late 1980s Casio boombox. In his tweed vest over plaid button-down shirt, thick-rimmed glasses, and artisan vintage tennis shoes found at Goodwill, he can often be found photographing the foam panda on his expensive artisan latte with an overly ambitious female friend he met on Instagram. Henry is a hipster. After hearing a new band on NPR's

Tiny Desk concert series, Henry buys the album on both vinyl and on Amazon music (iTunes is so cliché). But he is still not satisfied with the sound. So he converts the streaming Amazon content files into a loss-less format (to preserve their sonic authenticity), and then records some of those tracks in a better order (the band just didn't understand album design) onto a cassette. He voraciously and enthusiastically consumes this music through vintage headphones he found in his grandfather's attic and shares his accomplishment widely on his wide variety of social media accounts, offering up copies of his cassette tape to friends. He even has a few copies for sale at his weekly farmer's market table (at which he sells no food but instead farm-to-table knitted caps and pre-packaged organic cruelty-free granola mix that he made from things purchased at Whole Foods). The band's label hears news of the enthusiastic distribution of their recordings in this new form from several sources. The label sends Henry a letter: Cease and desist or we will sue.

Cases like that of Handlebar Henry bring up a range of questions: Was Henry morally justified in his audiophile-inspired file conversion? Do we hold him ethically accountable due to his bizarre clothing choices? As our case indicates, one major area affected by this property of distributedness is the commercial entertainment industry. From the early days of Napster to more modern examples of bittorrent file sharing and sites such as The Pirate Bay, there is a large community of Internet file sharers who deem the sharing of commercially produced digital content to be less morally reprehensible than stealing physical items from a physical store. However, this issue is complicated because these digital properties of reproducibility and transferability also ensure that it is much easier to download pirated media materials than it is to physically steal them from a store, particularly for young users familiar with the relevant technologies.

Professional ethics, as a reflection of societal norms and regulatory frameworks, play an important role in creative contributions, too. For example, what are one's obligations when remixing or "repurposing" content, a hallmark technique in what Henry Jenkins has called the "participatory culture" (Jenkins et al., 2009)? Both legal issues and ethical issues become complicated when copyright and intellectual property laws written for old forms of media are applied to newer and importantly different forms of media such as remixed, fan-produced YouTube videos and "original" MP3 songs that use the resampling of prior artists as primary source materials. Personal values play a role in an individual's decisions to download, alter, and/or share copyrighted materials; yet, these activities have also had significant implications for both social groups (e.g., college campuses which have implemented specific policies to combat electronic file sharing) as well as professional organizations (e.g., lawsuits filed by the Motion Picture Association of America). This chapter will highlight ethical issues of identity, agency, ownership, and privacy to better understand the digital-as-distributed. These issues are so important that we dedicate later chapters (specifically Chapters 7 and 8) to thinking through them in more specific detail.

Digital vs Analog Distribution

It is not the case, of course, that attributes like reproducibility and transferability are unique in and of themselves. Information has always moved beyond its port of call and been replicated, duplicated, and copied into various other forms and media. From the person-to-person tradition of oral cultures by memorization, to the text-to-person tradition of printed text cultures first replicated by scholarly hand and then later by means of the printing press, to the increased scale of sharing by photocopy and phonography, transfer and reproduction of knowledges has always been a ubiquitous part of organized human culture. The development and implementation of technologies have consistently shifted the means and methods by which these properties take shape. But, importantly, these technologies have all been *analog*. There is something importantly different about the reproducibility and transferability of information in *digital* form.

So what is that difference between analog and digital? When we are talking about analog and digital information, we are really talking in terms of signals: flows of information that are captured by interpretive systems (human, non-human, or mechanical). Most of the signals that shape the ways we regularly (or naturally?) experience the world are analog signals. Analog signals are continuous and infinitely variable. Digital signals are discrete and finite. Think, for example, of clocks. The gears driving an analog clock are in constant motion, progressing the second hand from one second to the next. The second indicator on a digital clock, on the other hand, is binary: It is either, say, 11:59:58, or 11:59:59. And this is true *no matter how far down you drill*. The most accurate digital clock imaginable is still either X, or Y, where x and y are some discrete fraction of a second. And if you could see it, the second hand on an analog clock can always be between two discrete numbers no matter how tiny they might be. This difference between continuous and discrete signals is fundamental to the difference between analog and digital sources.

Another way to conceive of this difference is by thinking of graphical representations of signals. When we look, for example, at the graph of a sine wave of an audio signal over time, we see a smooth change from positive to negative amplitude as the soundwave travels. Higher pitched sounds exhibit this sinusoidal change more quickly (the graph of that higher pitched sound is compressed). But the change is *continuous*: No matter how closely you examine such a graph, the signal is represented as a smooth curve. Interestingly, it is harder and harder to imagine this, since our experience of visualizing soundwaves, and maybe signals more generally, is done more and more through digital interfaces. And so we are more used to the practice of "zooming in" on graphs of signals to find that at some resolution, depending on the "quality" of the image, they appear to be discrete blocks of information. Digital signals reveal their discrete components in this level of visualization: The curve of the analog signal is made up of blocks, or bits, just like the visual pattern looks smooth at one level but is fundamentally a building block network of square bits. Again, this is a visualization of the difference between a continuous analog signal and a discrete digital signal. To contextualize this difference further, the ways in which a piece of music is encoded on analog media like cassette tape is very different from how that same piece is encoded in a digital medium. The analog encoding, as a continuous signal, allows for noise and variability, and the encoding degrades over time as the medium degrades or changes. The digital encoding, as a discrete signal, is perpetual since it is stored as discrete packages at some audio resolution (bitrate and sample frequency) such as MP3 or AAC standards.

Distribution and the Novelty of the Digital

Together, reproducibility and transferability define what it is for information to be *distributed*. Distributedness is a property of the digital in that what separates analog reproducibility and transferability from its analog forms are unprecedented changes in the speed and scale of information flow. Digital information moves both faster and at greater scales now than at any earlier point in human history—remember, as Moor (1997) tells us, it is "greased." This uncontrolled or uncontrollable flow of information has important normative implications that reflect its etymological roots in the Latin *distribuo*. *Distribuo* translates variably to divide, distribute, or assign. It is a term of categorization, or classification. You might take classification to be a structuring of the world, a giving-sense-to the ways in which the world works. The ancient Greek philosopher Aristotle took categorization as fundamental to his work, and gave us concepts like biological species, which are a reflection on the ways that living organisms differ. Yet any biologist will affirm that the concept of a "species" is a helpful heuristic but also problematic, with rough boundaries and irregularities coming to light with the more information we have about features of the natural world. If we once defined species by their form and function, we now throw genetic information, geophysical variability, and ecological interaction into that mix. So while *distribuo* denotes structure, it connotes disruption, too. The distributedness of the digital, as we see it, enables the unregulated and constant disruption of the shapes and flows of information.

This feature of distribution has at least two interesting implications to consider: one conceptual and one ethical. Consider that, on our conceptualization of distributedness as a function of either categorization or disruption of information, that property of the digital is not historically novel. Categorization and disruption are regular features of information flows and systems: changes to the concepts and frameworks through which human processes of knowing take place are constantly under renegotiation. Some of these renegotiations are radical departures from the norm—what Thomas Kuhn (1970) has marked as paradigm shifts—while others are subtle shifts in orientation to problems and ideas. Information has always been reproducible and transferable via various means, whether by verbal oration, written record, or press-printed copy—however, the reproducibility and transferability of digital information is uniquely marked by its *scope* and its *speed*. What is culturally delineated as the digital revolution is, on this reading, just a moment of significant change of speed and scope—but not something historically or ontologically novel. "The digital" is an always and already emerging phenomenon of human culture: the horizon of changes to the patterns of information distribution.

The other way to conceptualize the digital is a perhaps more standard view that the emergence of digital technologies, digital information, and digital devices marks a historically novel moment in human culture. The digital here is seen as marking a fundamental difference in the nature of information and in the epistemic and ethical roles it plays in the human experience, culturally and even biologically. A view like this parallels the physical conception of digital signals, as different in nature than analog signals. Recall that digital signals are discrete, fundamentally fragmented units of information: mere simulations of fundamentally continuous analog signals. So too, on this view, is the digital writ large a fragmented representation of an otherwise continuous analog. Framing the difference in this way opens up space for important critiques of the difference between the artificial and the natural, the machine and the organism.

The digital brings along with it a variety of ethical concerns in this altering the shape and texture of subjects with which it interfaces. In the next sections in this chapter, we address a series of subjects and cases as lenses through which to better understand distributedness as a property of the digital.

The Ethics of Distribution

In the first section of this chapter, we have offered up some intellectual grist for the mill to get you thinking about the nature of distribution as a property of the digital. In this section, we examine a series of issues and cases through which more closely to examine how distribution functions in context. Again, depending on the perspective one takes about the nature of the digital, the implications of its properties are either less or more far-reaching. We will consider what we see as the full range of its implications here, and leave it to you to determine the extent of the reach of the digital as distributed.

Since ethics is contingent on the nature of the world and our knowledge of it, consider first what it looks like to take the digital as transformative of the world and its relations. On such a view about the nature of the world, distributedness is a property that applies to all of reality. Such a view has been taken up most powerfully, perhaps, through the world of science fiction. Sci-fi can play an important role in philosophy in that it can enable us to gain some critical distance on our own experience and the see what it would be like for the world to be substantially different. We are thinking here of that famous scene from the Wachowski siblings' *The Matrix* in which Cypher decides whether to stay "plugged in." Cypher decides, ultimately, that he does not care whether the steak he is eating is *actually* steak or Matrix-steak, just so long as he *perceives it as steak*. The possibility for Cypher to be in the position of such a choice is made possible by his knowledge that there are, in this science fiction universe, two worlds. You and we do not have that luxury (as far as we know!) but this scene provides an important thought experiment, setting up conditions through which we can evaluate the possibility that "reality" is distributed across analog and digital worlds. It leads us to think through the idea that "the real" is not the static object that most of Western thought has historically taken it to be—upon or against which simulations and representations are evaluated—but, instead, a fluid and dynamic series of relationships that structure our perceptions of the world. Does that steak really represent steak? Does the social media profile you are viewing really represent a person? Does it matter? In what ways are identity and reality distributed, pirated, remixed, sampled, or transferred?

Relatedly, one might also think that distributedness is a property of the digital that applies to epistemological claims, or claims about what we can know. Epistemology, like all branches of philosophy, comes in many flavors, one of the most popular of which is that knowledge is constituted by justified true beliefs. To have knowledge, it is not sufficient just to believe X to be the case, but that your belief is justified in terms of its internal coherence (consistency with other beliefs you might hold) and external coherence (consistency with others' systems of beliefs). Your belief X must also be true; that is, measured against some relatively stable framework of knowledge. Some take truth to be a function only of what is known. Take, for example, the claim that Hillary Clinton did not win the 2016 U.S. presidential election: Given this true statement, nobody knows that Hillary Clinton won that election since "[o]ne can only know things that are true" (Ichikawa & Steup, 2017, para. 9). Yet not all truth claims work this way—the truth of some statement or claim is not contingent on our knowing it. Thus the "truth" condition offers us the connection between epistemological distribution and metaphysical distribution. "Truth is a *metaphysical,* as opposed to *epistemological,* notion: Truth is a matter of how things *are,* not how they can be *shown* to be" (Ichikawa & Steup, 2017, para. 11). If reality is distributed, then so is truth—and the important ethical implications of such a view are about our *access* to truth in the digital.

For example, epistemological distribution is easier to understand than its metaphysical counterpart, given the contemporary landscape of conversations about knowledge and truth. From climate change denialism to claims about fake news, and from Russian interference in elections to chat-bots passing for persons, claims to multiple "truths" muddy the conceptual waters in more frequent and diverse ways than perhaps ever before. And what happens when it becomes more difficult to justify and prove true our beliefs?

Unfortunately for justified true beliefs and fortunately for you reading this book, there are countless examples of the distribution of knowledge in the digital. One of the most newsworthy in the period we are writing this book is the relationship between Facebook and fake news. The March 2018 issue of *Wired Magazine* features an image of Facebook CEO Mark Zuckerberg, bruised and bloodied, representing the struggle he has faced over the previous two years. The interesting thing about this image is that it is fake. Its creator, New York City based artist Jake Rowland, composed it as a composite photo illustration. *Wired* reports that Rowland "mashed together an existing image of Zuckerberg with a photograph of a hired model—made up to look battered—whose features resembled that of the Facebook co-founder and CEO" (Wired Editors, 2018, para. 1). Rowland describes the final result of this composite as a "sort of a digital collage of these separate elements, a composite, that is a blend of fact and fiction, right down to the expression on his face" (Wired Editors, 2018, para. 3). The connotation of this image was not lost on its creator.

> I was also specifically inspired by the subject and our current moment. Those lines—between fact and fiction, document and forgery, what's "real" and "fake"—have never been so slippery and difficult to detect. The manipulation of information and imagery online, especially via social media, and in digital media generally, is being used to warp our perception (and influence our politics) in unprecedented ways. The digital is spilling over into the physical world at an accelerating rate in glaringly obvious ways—and in ways we probably can't even consciously perceive yet.
>
> *(Wired Editors, 2018, para. 5)*

Rowland's perception of the distributedness of the digital is oriented to *ontological* and *epistemological* issues—problems of reality and problems of knowledge. These sorts of issues do not tend to stick with or bother those in industry, unless they result in applied problems with significant value-laden implications. Put otherwise: Non-philosophers have a hard time seeing the metaphysics of the forest apart from the ethical impact of the trees. Rowland is keyed into the possibility of problems "we probably can't even consciously perceive yet" precisely because he is looking back at a year of problems caused by the spillover of the digital made very visible. Zuckerberg has held to a tradition of self-improvement since 2009, holding himself publicly accountable to a New Year's resolution each year. In 2018, that resolution was to fix Facebook (Zuckerberg, 2018). In a January 4, 2018 Facebook post, he wrote: 'Facebook has a lot of work to do—whether it's protecting our community from abuse and hate, defending against interference by nation states, or making sure that time spent on Facebook is time well spent. My personal challenge for 2018 is to focus on fixing these important issues" (Zuckerberg, 2018, para. 4). Later in that same introspective post, he analyzed the heart of Facebook's problems in terms of power: the ability of digital technologies to decentralize or centralize power. "Back in the 1990s and 2000s," he writes, "most people believed technology would be a decentralizing force. But today, many people have lost faith in that promise. With the rise of a small number of big tech companies—and governments using technology to watch their citizens—many people now believe technology only centralizes power rather than decentralizes it" (Zuckerberg, 2018, para. 7). Implicit in this statement is the relationship between power and knowledge—a relationship at the heart of Facebook's controversy. Zuckerberg's resolution to fix Facebook is personal because it compels him to rethink the way he views the role of his digital technology, to rethink the extent to which Facebook is a mere platform versus a content provider, each of which comes with its own landscape of ethical responsibilities to support informational liberty or restrict information. In a March 2018 interview, Zuckerberg noted, "A lot of the most sensitive issues that we face today are conflicts between real values, right? Freedom of speech and hate speech and offensive content. Where is the line?" Journalists noted that he sounded here "more like an ethics student than the billionaire CEO of one of the world's most valuable and influential companies" (Swisher & Wagner, 2018). Zuckerberg and his employees learned that their open and neutral digital platform, with its mission to empower people, was being used to suppress people, to bend messages, to manipulate news, and to influence politics (see Wired, 2018). The revelation that Russian trolls and bots were bending news and having a growing practical influence on American social and political dynamics brought these issues to a head. It was so much more than a neutral digital force—its nature to distribute information in increasingly fast and diverse ways—had taken on a new form of agency.

Agency as Distributed

In the digital, epistemic distribution is affected through the distribution of agency. In analog modes of existence (think here particularly of the Western Enlightenment), agency was taken to be an internal and rational function of the human being—something that could be lost with failures of rationality and problems with internal functioning of the body and mind (see Chapter 8 for our development of agency in digital ethics). In many non-Western or Western minority views, agency still privileges the human, even while recognizing the powerful role of social embeddedness and community. In the digital, this modernist view of agency is threatened by a fast and vast expansion of modes of action.

N.K. Baym, in work on personal connections in the digital age (2010), writes of this expansion in terms of the "reach" of digital information—an idea that strongly mirrors Moor's *greasedness*. From this reach of information, Baym argues that

> Much of our mediated interaction is with people we know face-to-face, some media convey very little information about the identities of those with whom we are communicating. In some circumstances, this renders people anonymous, leading to both opportunity and terror. In lean media [a term referencing the information-carrying capacity of a medium (Baym, 2010,

p. 60)], people have more ability to expand, manipulate, multiply, and distort the identities they present to others. The paucity of personal and social identity cues can also make people feel safer, and thus create an environment in which they are more honest.

(Baym, 2010, p. 9)

The landscape of identity to which Baym points is also a landscape of agency: Persons (and we are being careful here in acknowledgement of the potential for both animal and artificial intelligences) can not only present themselves in different ways thanks to the expansiveness of information and venues for communication, but those personae can each have causal consequences in the world. A reasonable interaction from Donald-in-person might clash in consistency with a vicious tweet from Donald-on-twitter, and those both might be contradicted by Donald-as-anonymous-Redditor all within a tight time frame. Do we count those as the same agent acting in the world? The modern philosophical mind would only have recourse to attributing the agency to Donald, who is seen as having developed a novel form of digital schizophrenia. Yet the digital philosophical mind might see Donald-variants as each acting in the world in their own ways, supported by networks of reactions and interactions that take on an agential life of their own, independent of the original intent of Donald-in-person.

And agency is not only mediated by technology but also given over to it in important ways. An example of this technologically mediated distribution of agency is the February 5, 2018 stock market drop of over 1,400 points. That drop was not only the largest single day drop in U.S. history, but it also happened over a mere 10 minutes, more quickly than any prior stock market loss (Egan, 2018). This novel event was made possible by the distribution of agency from stockholders through stock market floor traders to supercomputers, programmed by human agents to algorithmically react to fluctuations in market forces far more quickly than humans ever could. The digitally distributed exchange between two of these computational nonhuman agents led to a run-away sell-off with at least short-term economic implications. As David Berry thoughtfully points out in his 2016 *The Philosophy of Software*, "[o]f course, we have always used devices, mechanical or otherwise, to manage our existence, however, within the realm of digital computational devices we increasingly find symbolically sophisticated actors that are non-human" (2016, p. 125). This extension of human agency to agential computational machines, designed and built by human agents, is the material for popular science fiction not because it is fictional but because it has growing technological social, economic, political, and ethical implications in the digital.

Ownership and Privacy as Distributed

A second ethically relevant implication of the distributedness of the digital is its impact on questions of ownership and privacy. The ownership of information and the capacities by which those owners can choose to maintain that information as private is fundamentally different in the digital than in analog modes of existence. While we can participate in information exchanges in impressively diverse ways, that same diversity pushes the ethical norms about what it means to own or keep private information. Individuals have a legal right to privacy if they can reasonably expect their information exchanges to be private. This legal right has been articulated by the use of an analog analogy of a house: If you are in an argument with a partner in your home, with your windows closed, you have a reasonable expectation of privacy. If someone on the street outside were to somehow overhear your argument and use it to extort or bring some other legal case against you, you would be in a strong position to argue that the information was private. However, if you are having that same argument at home, but with the windows open—your legal case is weakened since that same person on the street could easily overhear your argument and your expectation of privacy was far less reasonable. In the argument of this section to follow, we argue that the distributedness of the digital opens windows, even when we want them to be closed.

Participation in flows of information continues to increase in speed and depth, driven in part by distributed access to that information. With analog sources of information, the source and the interpreter of that source had to be in shared physical space. The idea of curling up with a book, for example, might bring to the imagination of the analog reader pictures of holding their favorite paperback in one hand, with fingers spread across the fold to support the spine, while lying on their sofa in a living room on some rainy afternoon. That reader and that content of the paperback were connected by the physicality of the book itself. In the digital, access to the information contained within that same paperback book is distributed across platforms, devices, and formats in myriad ways. And the difference between analog-only and analog–digital book content is radically decreasing, with continued efforts by content-providers like Google and publishing houses working to convert analog information into digital form. The ubiquity of mobile computing means that more often than not, we end users have the ability to access the same information from a widely distributed network of sources, in multiple forms, across platforms and devices. The same book that readers could have accessed only in analog form when it was first printed, say, in 1980, is now available in e-book formats, as a digital audio book, as a PDF, through libraries both physically and digitally, and in snippets through preview sources like Google Books. Researchers and readers who once, were they so inclined, had to take notes on paper cards to remember references and quotations, can now snap a photo on their smartphones and upload to the cloud, to be referenced from anywhere at any time. You can curl up on the couch with the contents of book, but without the book, in a still-increasing diversity of ways.

This same distributedness of information has already demonstrated significant and novel ethical implications. To offer just a few examples, consider first the phenomenon of bitcoin mining. At the height of the craze of that cryptocurrency, "miners" were utilizing every bit of processing power they could to unearth new bitcoins by linking together validated transactions. One ingenious and potentially nefarious solution is to gain a leg up on other miners by utilizing other people's processing power: Bitcoin-mining spyware is one of the largest early 21st century threats to cybersecurity (Condliffe, 2018). The same strategy employed here was employed in the late 20th century for less nefarious scientific purposes in the SETI@home program, which utilized an early form of crowd-computing in their processing and data intensive search for extra-terrestrial life. In the case of SETI@home (The University of California, 2018), however, the end user had to volunteer their computational power to the project by download-ing and running software on their end. These two examples, of an intuitively "bad" and an intuitively "good" use of distributed computation of distributed information, are only the tip of iceberg here.

Distributedness of the digital has allowed for the growth of big data and citi-zen science initiatives on the empirical/scientific end of things, but also to the possibility of the Arab Spring of 2010 through the explosion of social media forms of information exchange. Yet, "digital permanence" also presents a novel vulnerability here in ways that the permanence of analog information has not been: Both the sheer amount of information exchanged and the types of media through which the exchange happens bring up complex problems of storage and retention. The so-called Information Explosion, the ever-increasing growth in the amount of information in the digital, means that more and more of our human cultural knowledge, as well as our empirical and relational knowledge of the world is contingent upon the successful storage and exchange of it in digital form. Persistence empowers participation by capturing the history of informa-tion exchanges in support of change over time. By enabling the participation in discourses and information exchanges anywhere, in multiple forms, and across multiple platforms and devices, the distributedness of the digital both empowers and makes vulnerable individuals and communities in ways different from previ-ous modes of information exchange.

Three Examples

We will offer three very different examples here to get us thinking about ethical concerns around distributedness of the digital. The first is a well-known story that emerged with the radical departure from music distributed on media (think everything from 8-tracks to cassettes to compact disks to vinyl). In the early 2000s, the RIAA, or Recording Industries Association of America, made it their legal and moral quest to sue the pants off of a select few college students who were sharing music files on the high-speed networks on some college campuses. As of 2006, when the RIAA stopped reporting the number of copyright infringement lawsuits filed, their count of lawsuits filed stood at 17,587 (Collegestats.org). The industry argument was indeed both legal and moral: They claimed that their clients had a legal right to the sale and distribution of their music (at the time primarily on compact disk) and that these student file-sharers were breaking the law. But they also argued that sharing files was morally wrong: that denying others the ownership of their sonic property was no different than stealing a car, breaking into a house, or robbing a bank.

Remnants of this argument remain even now, in the form of the anti-pirating messages often shown at the start of feature films in cinemas. They claimed that the continued and escalating distribution of digital music outside of constrained physical media like CDs would ruin artists' economic incentives to make music, and destroy the very thing that students were so interested in sharing in the first place. By keeping music contained to CDs, owners of that music could control its distribution and maintain economic solvency—and, of course, keep making music. Since that short period of legal flurry, both the public and the recording industry have come to learn that, in fact, digital distribution of music media forms actually has *increased* participation in music consumption (Aguiar & Martens, 2016). And, more importantly, the overall failure of this legal drama to support RIAA's claim evidenced an important moral statement: people simply do not care about copyright laws as they apply to digital music (Collegstats.org). The distributedness of the digital shifted the grounds of the ways we make, listen to, share, purchase, replicate, and engage with music and bands. And it radically shook up the analog recording industry.

A second example is the mid 2017 Equifax data breach. Equifax, one of the U.S.'s three large credit monitoring corporations, acknowledged a massive scale data breach back in 2017 (FTC.gov). Millions of Americans had their personal information stolen, unable to be tracked, and with the potential for free and wide distribution. Breaches of privacy are not new: Indeed, they have happened regularly at various scales throughout history and continue to happen now. However, the notable comparison here is to a breach of physical documents versus digital ones. If thieves broke into an insurance company's office even in the early 1990s, they would have the potential to steal file boxes full of paper that contained personal and private information about clients. They could have then perhaps physically copied the files and distributed them to criminal organizations worldwide. Yet the scale and speed of this sharing is dwarfed by the potential cyber-thieves have to distribute stolen information in digital form. It is harder to imagine that thieves would even find it worth their time to steal paper files, if their goal was amount of data rather than specific datum: They would be foolish thieves, indeed. The Equifax breach made national headlines thanks to its scale, but cyber-breaches like these are all too common. In both cases, the response was—almost necessarily—tepid: The organizations did not know how or exactly when the breach had occurred, exactly why thieves wanted it or what they intended to do with it, where it had gone, or how best to protect their community members in the future. The offer of identity-theft monitoring was the only recompense that could be offered, despite the potential harms extending, given the nature of digital information, potentially in perpetuity.

As a third example of the ethical implications of the distributedness of the digital, consider the debates around ownership and privacy of genetic information. This debate has been driven most publicly by the recognition of the source of the infamous HeLa cell line, a line of cells that have the remarkable ability to survive and reproduce outside of the human body seemingly infinitely and immortally. HeLa cell cultures have played a role in almost every human biological breakthrough of the late 20th century, and have been a staple of biological laboratories since their discovery in 1951, when Ms. Henrietta Lacks walked into the Johns Hopkins hospital for biopsy of a cervical tumor. Yet HeLa cells remained unnamed and unclaimed until 2011, when journalist Rebecca Skloot (2010) connected historical dots and tracked down Ms. Lacks' family. Since that time (so read here, remarkably recently) and in conjunction with continued advancements of our understanding of genetic information, we have begun to unpack and analyze ethical tensions around who has the right to genetic information and to what extent that information should and can remain private. Biobanks, those sources of aggregated genetic information, were arguably safehouses of de-identified and/or anonymous genetic information researchers began to discover a "spectrum of possibilities" based on developing understanding of genetic variance along which biobanked samples might be re-identified (Gutman, 2013). Now, from the genetic information contained on a single cheek swab, others can determine your cultural heritage, phenotypical traits like eye color and hair color, your risk for particular disease, and even what your face looks like (Aldhous, 2014). It can also identify your relatives, even if they have never shared their own genetic information (Murphy, 2018). In its analog form, your genetic information remains hidden, coded in the DNA and RNA relationships within your physical form. But when converted or interpreted digitally, outside of the body and through computational methods, your information becomes visible—and distributed. What to do with this potential risks and threats to privacy here is nothing we have yet agreed on. In 2017, the medical and scientific communities held a period of open commentary on what is known as the "Common Rule," or the consensus of the research community on how to protect human individuals in research. As a result, the leading committee of the latest revision to the Rule considered including digitized information from biospecimens as something to be protected—but in the end they did not (Menikoff, Kaneshiro, & Pritchard, 2017). But the shift from physical to digital genetic information has opened the conversation to consideration of the human subject as something more than physical—as a community of data that is distributed across physical and digital media and persisting over a new scale of time and space.

Responsibility as Distributed

As a reminder, there are several ways to conceive of the digital: We have presented polar positions of the digital as either technological shift or change in worldview. In this chapter we have argued that the digital is not just a paradigm of emerging technologies or a mere technological shift toward binary information structures driven by those computational technologies. Instead, we suggest it is a transformation of the nature of the world and the ways human agents come to know and act in it.

Ethical implications are often considered in terms of the results of certain conduct: Does distributedness cause harms or offer benefits in novel ways that we ought to more carefully consider? This is, you will recall from Chapter 2, the important focus on questions of character, a driving force of consequentialist ethical reasoning. But on the other end of the normative spectrum, the question of responsibility takes center stage. And, similar to our analysis of the distributed consequences of actions, we see responsibility as something importantly distributed in the digital, too. It is contingent on agency, the ability to act, on epistemology, and on context or case. It is made possible through digital technologies, affected by the digital divide (social-economic barriers to information), and relevant to questions of marginalization and cultural expectations.

These attributes of distributedness, as concepts, carry about with them normative weight. We still may be more apt to give credence to scientific knowledge when the data on which it is based is reproducible. And reproducible claims are more important when they can be transferred between contexts; that is, when knowledge of x helps us better understand y, and z, too. Epistemic reliability is grounded on reproducible results. Thus, by analogy at least, digital information (as distributed) is taken to be relevant, or pragmatically true, when it is constituted by reliable signals that are easily replicated. The claim that we have just laid out is a loaded one. Consider what accepting it entails: Any time you have a reproducible and transferable signal, you've got some claim to true information. Yet this flies in the face of other implications of the digital; namely, that the digital facilitates the simulation of information - not its copies but its identical replication, a process that opens space for what philosopher Jean Baudrillard critiqued as the simulacra, or empty representations. Hence, the good begets the bad. The ethical implications of this position point to the ways in which the digital denies the binary normative evaluation of goodness and badness, and pushes us to reconsider how to make sense of this idea of digital ethics.

Next Up

In the next chapter, we look at moral algorithms and machine ethics as a lens through which to unpack programmability and procedurality as key properties of the digital. The binary nature of digital information structures not only the programmability of digital media but also its procedurality. We trace a conceptual genealogy to show how it challenges traditional ethical problems that have been grounded in ontological conceptions of the uniqueness of media objects. We then apply our analysis to some of those contemporary issues, including advanced robotics.

References

Aguiar, L., & Martens, B. (2016). Digital music consumption on the internet: Evidence from clickstream data. *Information Economics and Policy, 34*, 27–43.

Aldhous, P. (2014). Genetic mugshot recreates faces from nothing but DNA. *New Scientist.* Retrieved from https://www.newscientist.com/article/mg22129613-600-genetic-mugshot-recreates-faces-from-nothing-but-dna/.

Baym, N.K. (2010). *Personal connections in the digital age.* New York: Polity Press.

Berry, D. (2016). *The philosophy of software: Code and mediation in the digital age.* New York: Springer.

Collegestats.org. (2018). RIAA vs. college students: File sharing lawsuit statistics (infographic). Retrieved from https://collegestats.org/2010/02/the-riaa-vs-college-students/.

Condliffe, J. (2018). Forget viruses or spyware – Your biggest cyberthreat is greedy currency miners. Retrieved from https://www.technologyreview.com/s/609975/for get-viruses-or-spywareyour-biggest-cyberthreat-is-greedy-cryptocurrency-miners/.

Egan, M. (2018, February 5). Dow plunges 1,175 - worst point decline in history. *CNNBusiness.* Retrieved from https://money.cnn.com/2018/02/05/investing/st ock-market-today-dow-jones/index.html.

FTC.gov. (2018). The Equifax data breach. Retrieved from https://www.ftc.gov/ equifax-data-breach.

Gutman, A. (2013). Data re-identification: Prioritize privacy. *Science, 339*(6123), 1032.

Ichikawa, J.J., & Steup, M. (2017). The analysis of knowledge. In Edward N. Zalta (Ed.), *The Stanford encyclopedia of philosophy. (Fall 2017 Edition).* Retrieved from https://plato. stanford.edu/archives/fall2017/entries/knowledge-analysis/.

Jenkins, H., R. Purushotma, M. Weigel, K. Clinton, & Robison, A.J. (2009). *Confronting the challenges of participatory culture: Media education for the 21st century.* The John D. and Catherine T. MacArthur Foundation reports on digital media and learning. Cambridge: MIT Press.

Kuhn, T. (1970). *The structure of scientific revolutions* (2nd ed.). Chicago, IL: University of Chicago Press.

Menikoff, J., Kaneshiro, J., & Pritchard, I. (2017). The common rule, updated. *New England Journal of Medicine, 376*(7), 613–615.

Moor, J. (1997). Towards a theory of privacy in the information age. *Computers and Society, 27*(3), 27–32.

Murphy, H. (2018, October 11). Most white Americans' DNA can be identified through genealogy databases. *The New York Times.* Retrieved from https://www.nytimes. com/2018/10/11/science/science-genetic-genealogy-study.html.

Skloot, R. (2010). *The immortal life of Henrietta Lacks.* New York: Broadway Paperbacks.

Swisher, K. & Wagner, K. (2018). Mark Zuckerberg says he's 'open' to testifying to Congress, fixes will cost 'many millions' and he "feels really bad'. *Recode.* Retrieved from https://www.recode.net/2018/3/21/17149964/facebook-ceo-mark-zuckerberg -congress-data-privacy-cambridge-analytica.

Thompson, N. & Vogelstein, F. (2018). Inside the two years that shook Facebook – and the world. Retrieved from https://www.wired.com/story/inside-facebook-mark-zuckerberg-2-years-of-hell/.

The University of California. (2018). SETI@home. Retrieved from https://setiathome. berkeley.edu.

Wired Editors. (2018). What happened to Zuckerberg? Here's how our March 2018 cover was created. Retrieved from https://www.wired.com/story/facebook-about-the-cover/.

Zuckerberg, M. (2018, January 4). No title. Retrieved from https://www.facebook.com/ zuck/posts/10104380170714571.

5

MORAL ALGORITHMS AND ETHICAL MACHINES (PROGRAMMABILITY AND PROCEDURALITY)

What does it mean, ethically speaking, to have programmable instructions as core features within a technology? To answer this question, we will consider how software algorithms are expressed through programs and procedures. Although their logical, rule-based composition may seem to leave little room for ethical consideration or debate, algorithms reflect the values of their human designers and offer a rich source of analysis for considering the ethics of software (and, ultimately, of the digital devices that run this software). Additionally, algorithms present opportunities to make or receive arguments interactively and procedurally, ideas that have been extensively discussed in relation to videogames and other types of interactive digital media. Moral algorithms, or algorithms with ethical capabilities, are critical for autonomous software that makes decisions without human intervention or oversight. We will explore each of these avenues and topics within this chapter. For now, though, let us start with an example of algorithms at work. Consider the following case of a seemingly harmless leisure activity.

CASE STUDY: THE UNREAL ELDERLY

In the summer of 2019, social media enthusiasts on Facebook and Twitter delighted in uploading and sharing artificially aged photographs of themselves and family members. The doctored photographs were produced by facial recognition software with computational techniques for manipulating imagery to provide the illusion of age (e.g., adding wrinkles or sun spots, removing hair or providing a receding hairline, and so on).

While the activity was a fun diversion for many who tinkered with these mobile applications and shared the results through their social media accounts, the process also raised some serious ethical concerns surrounding one's privacy and one's personal photographs (and what happens to these photos when they are uploaded to an unknown server hosted by unknown companies). These artificial aging photography tools were the latest examples following a trend of algorithmically enhanced audio/video software (other notable examples include "deepfakes," or audio/video of real people saying or doing things they never actually said or did (Chesney & Citron, 2018) and a web site that uses AI and machine learning to generate photorealistic images of people who don't actually exist and yet are indistinguishable from real human beings (see https://thispersondoesnotexist.com/).

Privacy issues aside, several other moral complications surround these algorithmic representations. For example, returning to our artificial aging example, what does this activity suggest about how younger people see the elderly, and how might older adults view the trivialization of growing older and experiencing the various physiological and psychological changes that accompany the aging process? As with many of the examples we discuss in this book, a fundamental issue here is the difference between an *algorithmic* process, or a process executed by computer software, and a *biological* process, or a process experienced by an organic entity such as a human being.

Algorithmic and biological processes are not equivalent, despite the propensity for many to equate parts of the human body to machinery, such as the metaphor of the computer as a digital brain. There are obvious differences in composition between bodies and computers. A body includes cells, organs, tissue, neurons, and blood and a machine is made of materials such as silicon, plastic, fans, and wires. However, even beyond the obvious differences in material composition, these processes are quite different in function and capability. For example, algorithmic processes take place on a completely different scale of time than biological processes. Computers "work" much faster, hence the ability to quickly create very realistic looking—but completely artificial—photographs of individuals that far outpace their bodies' natural aging processes.

Despite their material and functional differences, however, there is a moral relationship between human designers and their digital computers. Most algorithms are authored by humans, introducing the possibility of human bias and error into the instructions followed by the algorithms. (Some branches of computer science are experimenting with algorithms that write other algorithms, but even in this case, a human wrote the original algorithm somewhere down the line.) Thus, algorithms may be evaluated as moral if authored in an ethically responsible fashion and evaluated according to an ethical framework, or immoral if designed in a less thoughtful way. In the case of these age-based photo manipulations, an algorithmic process using machine learning allowed software to manipulate existing photos to simulate what individuals might look like later in their lives. The biological authenticity of such representations may take many years to verify, depending on one's initial age when using the application. The moral dimensions of such an algorithm, however, can be debated here and now.

What, then, are the broader ethical implications of algorithms beyond what we have discussed here? In this chapter, we investigate algorithms more carefully from an ethical perspective by drilling down into two specific aspects of the digital and its software: programmability and procedurality.

Programmability and Algorithmic Bias

In Chapter 4 we discussed the properties of digital movement between sources— that is, the ability of digital data to be reproduced, transferred, and distributed across vast networks. Now we consider movement within software itself. By this we mean the way that digital data can be transformed, manipulated, "executed," or "run" as code. This is another fundamental difference between analog and digital data. Analog data exists as it is once it is created. In terms of its material composition, it is static. But digital data is not. It is dynamic in that it can be transformed, extended, manipulated, compressed, encrypted, and reshaped. Through certain algorithms, it can even change itself over time, in a generational sense, through machine learning or genetic algorithms. And all these potential manipulations are possible due to the programmability of digital information and its media forms.

Programmability means that digital computing devices are constructed to follow predefined instructions in order to function. They are *programmed* to perform differently depending on the specific directives they are built to follow and the order and fashion in which those instructions are embedded into computer software. Programming can be done in many different computer programming languages, from so-called "high-level" computer programming languages such as C++, Java, or C#, to "low-level" computer programming languages like Assembly. High-level programming languages are written with more natural language constructs and are easier to understand and use. However, that usability comes at a cost since low-level languages may run more quickly and efficiently since there is less abstraction between the code and the instructions processed by a CPU.

When programming rules are lumped together with a particular purpose in mind, they are called algorithms. These algorithms can be expressed in many ways—as computer programs, as logic tables, as pseudocode, or they can be described in a "natural" language like English, Spanish, or French. Because programmable algorithms are useful for so many different applications, there are many different types that have been developed for common tasks in computer software such as sorting, compression, and encryption. Algorithms are often compared to the rules of a recipe: Given a set of ingredients, the algorithm is the set of steps required to turn those ingredients into some new form (hopefully edible).

Given those same ingredients, the same algorithm should yield the same results every time. Only by varying the type or quality of the ingredients can a different meal be created at the end of the process.

Science fiction authors regularly feature algorithms for printable food in their fictional world building—and we've certainly seen this concept presented to us in fictional television shows like *Star Trek*—so the concept of edible ingredients as data input is not necessarily out of the question for our future computer programs. However, instead of edible ingredients, most of our current computer programs use less exotic types of input data such as numbers, characters, and strings of text. These data can be manipulated by the sequence of steps that make up the algorithm. For example, a simple algorithm to calculate sales tax on a purchase would use different types of numerical data to hold values for the purchase price of the items, the sales tax rate, the calculated sales tax, and then the overall cost of the purchase. Fancier algorithms might also include certain types of error detection, such as making sure that the purchase price for each individual item is within a certain range (perhaps above zero and equal to or below the highest price item in the store). A series of steps within that algorithm then operate on that data in a step-by-step fashion until the final result, or output, is calculated. A very simple three-step version (without error detection) of the overall algorithm might look something like this:

1. Calculate the subtotal by adding together the purchase price of each individual item.
2. Calculate the tax by multiplying the result of step one by the sales tax rate.
3. Calculate the grand total by adding the subtotal to the tax.

This sales tax algorithm is very basic, of course, and would need to be turned into a program in order to be used. This would involve a programmer implementing this algorithm using a specific programming language, like Java or C#. While undergoing this process, the programmer would likely think of some additional tweaks to make to the algorithm. The algorithm might now need to include some error checking, such as to ensure that the purchase price of each item is between a certain allowable range, such as *zero* and *maximum price*. The algorithm would also need to ensure that common mathematical problems, like attempting to divide by zero, are screened out through error checking. When these types of errors are encountered, the program could present an error message to the user or it could simply stop executing the steps in the program and halt.

It is important to discuss programmability and algorithms when talking about the ethics of digital technology because our digital machines can be thought of as implementations of algorithms (Moschovakis, 2001). We can think about the ethical implications of this relationship from several perspectives, but one angle to consider is the seeming infallibility of these logical devices. Many people are inclined to naturally trust computers, because they are machines and machines are often characterized as logical and objective devices that do not make mistakes. This is untrue because while algorithms and the computer programs that "execute" them may indeed be very reliable (assuming they are operating with normal parameters with no hardware failures or environmental problems), they are still designed by imperfect creators, as we discussed previously. Since computers are usually programmed by humans, the same ethical problems that a human might encounter in conversation or action can also be found in computing technologies when biased code is written and run by programs. Moral algorithms, in other words, must be produced through careful thought and intentional design, not only in regard to the specifications of a program, but also in carefully considering that program's use (e.g., functionality and operation) and context (e.g., stakeholders, access, equity, etc.). It may seem strange to think that computer software can be biased, morally problematic, or even immoral, but there are many examples of bias found in the hardware and software systems used in different types of algorithms, from basic calculators to sophisticated image detection programs.

In the sales tax example above, for example, a programmer might miss common types of errors simply because of his or her socioeconomic status, background experiences, ethnicity, implicit biases, ideological positioning, or (correct or incorrect) assumptions about the program's intended use. The context of the program's use is important to consider here. A program written to run on a cash register in a local bodega, for example, might require different accounting and algorithmic considerations than one written for a multinational bank. This is due to both the differences in the amounts, numbers, and scale of transactions as well as the differences in policies, procedures, and stakeholders involved with each location. Unless the programmer implementing the register algorithm has deep experiences visiting both bodegas and multinational banks, it is entirely reasonable to expect that some important user scenarios might be missed if one program is intended to serve both types of locations.

Other complications arise from more complex algorithms and the way they are developed. For example, some early and widely used programming code that was used for face detection had more difficulty recognizing darker-skinned individuals than lighter-skinned people. In her work with an open source software package, one graduate student researcher at the MIT Media Lab noted that "wearing a white mask worked better than using my actual face" (Tucker, 2017, para. 4). Because the algorithms worked using stored data sets of digitally encoded skin tones as benchmarks, they lacked the necessary diversity to identify a broader range of faces and skin tones. Further, because the programming code was an open source data set, it was used widely in a variety of different applications and this bias was replicated across many different systems that relied on facial recognition to function. As a result of this experience, this MIT graduate student researcher, Joy Buolamwini, used the experience to motivate her research on "the coded gaze," a line of work that investigates how to address bias within algorithms and computer programs (Buolamwini, 2016). Other work has found bias in the algorithms used in popular search engines such as Google. For example, researcher Safiya Noble (2018) found that when searching for "black girls" using Google's algorithms, advertising results were both hypersexualized and pornographic. Noting the lack of substantial ethical training in most mainstream engineering programs, she asks, "If Google software engineers are not responsible for the design of their algorithms, then who is?" (loc. 1181).

Programming devices also means squeezing ideas into digital boxes (programmatic structures) that do not necessarily align with the broader contexts of those ideas. Wachter-Boettcher (2017) recounts a traumatic experience from when she was seeking a birth control refill and was forced to answer a "yes" or "no" question about prior sexual assault on a PDF form in order to register as a new patient at a clinic. The rigidity of the form reminded her of the many prior problems she had experienced with digital forms: sign-up forms that demanded to know one's gender, with only options for male and female; college application forms that assumed the applicant's parents lived together; and race and ethnicity information requested without the option to select multiple races or ethnicities. For the programmers who designed such forms and software programs, these real-world scenarios were probably simplified so that a computer could ultimately convert things to numbers and operate upon the data accordingly. The problem, of course, is that real life is more complex than this. Parents of college students get divorced. Individuals identify as transgender or choose a nonbinary category for representing their gender. People celebrate a mixed racial heritage and draw from multiple ethnic traditions. By breaking down information into a format that is easier for computers to understand, we sometimes also adversely affect the humans using the computers. For Wachter-Boettcher who was just seeking some birth control pills, the form's approach to soliciting information about sexual assault was uncomfortable and artificial.

Many other examples of algorithmic bias are found in the everyday digital systems with which we interact. For example, in a *New York Times* opinion piece entitled "Artificial Intelligence's White Guy Problem," Crawford (2016) recounts many examples of software bias due to sexism, racism, or other forms of discrimination. Specific examples included a photo application from Google where black people were classified as gorillas, software from Nikon that incorrectly classified photos of Asian people as blinking, a same day delivery service from Amazon that excluded zip codes from predominantly black neighborhoods, and an algorithm from Google where women were less likely than men to see advertisements for jobs with high salaries. Many of these algorithms were based on models of artificial intelligence implemented through machine learning. As Crawford noted in her analysis, "artificial intelligence will reflect the values of its creators" (2016, para. 14).

There are other examples we could consider where algorithms seem to *remove* discriminatory practices by taking humans, and their implicit or explicit biases, out of the equation. For instance, a *C/NET* article published in October of 2018 recounts an African-American woman's experience shopping in a recently opened Amazon Go cashierless grocery store. Author Ashlee Thompson (2018) describes her new (and unfamiliar) approach to shopping in an environment in which digital cameras monitored the environment rather than one or more human employees. She describes how she was informed of the "rules of shopping" as a child growing up—rules like not touching anything, keeping her hands in her pocket, and not digging through her purse (when she was old enough to carry one) (2018, para. 1). In contrast, when shopping at Amazon Go, there was no human person monitoring her presence in the store—only cameras and algorithms to determine what she placed in her bag so that those items could be charged to the payment method on file in her Amazon.com account. She summarized her thoughts shopping in this new digital environment at the end of the transaction like this: "No one cared what I was doing. Is this what it feels like to shop when you're not black?" (para. 18).

It should not escape our notice that although Thompson *feels* like she is less monitored when doing her shopping in this Amazon Go store, she is actually being monitored *more closely than ever before*, but by digital technology rather than humans. In this case, though, the technology seems safer than the human cashiers who the author has previously encountered. And that is a sad thing, in some ways, but it is also a hopeful example of the power and potential of digital technology to level the playing field, or, in this particular case, the grocery store aisles.

Another implication of the programmability of our computers is that they are designed to be normative. What this means is that there is a "normal" or "standard" output that can be evaluated for accuracy and precision as the technology is used. As Moor explains:

> By its nature, computing technology is normative. We expect programs, when executed, to proceed toward some objective—for example, to correctly compute our income taxes or keep an airplane on course. Their intended purpose serves as a norm for evaluation—that is, we assess how well the computer program calculates the tax or guides the airplane.
>
> *(Moor, 2006, p. 18)*

As we think about programmed digital technologies, then, we know that the programs and algorithms they use are normative and are working toward a "right" answer. As MIT researcher Joy Buolamwini proved, however, even when an algorithm performs well in reaching that right answer, if other data components are missing or not fully representative of the stakeholders who use the technologies, then the evaluation may fail (Buolamwini, 2016). "Garbage in, garbage out" is a famous phrase in computer science. This means that if bad data is provided as an input, then you can expect to see bad data at the output, too, regardless of how efficient and foolproof you have designed your algorithm to be.

Further, programs may work well on one level but provide disappointing results when benchmarked against another, more unexpected or organic, type of outcome. The benchmarks used to evaluate normative conditions may also have emerged from a set of values different from those of the individuals using the software or hardware devices. For example, if a business CEO outsources a piece of software to a development company, she may value efficiency, profit, and standardization. The software developer is likely to use those standards as guidelines during the creation and testing of the product. The end-users of the product, however, may pay a price in terms of the product's overall usability, their workplace privacy, or other factors that were not deemed as essential to the product as it was developed. Yet, the product is still seen as successful by the client because it met the normative expectations of the CEO.

Another notable aspect of the programmability of our digital devices is that they are extensible, or able to be extended in new directions. This means that although they may start out by following one set of programmed instructions, the algorithms employed by their designers may allow the programming rules to adapt and change depending on human or other factors (e.g., chronological, environmental, etc.) We commonly see this in mobile computing devices, for example, where programmed instructions specify the general parameters for control panels and other aspects of the interface, but end users can customize or personalize the code to be more responsive to their own individual needs. Consider again the Amazon Echo, originally developed to be a voice assistive device for helping with routine household matters (e.g., checking the time, doing conversion rates for measurement tasks while cooking, playing music, or adding items to a grocery shopping list). By talking with the device, nicknamed Alexa, users could provide instructions to a computer that would filter the voice results into algorithms that would allow the device to respond accordingly. Over time, however, the Amazon developers programmed the technology so that it could evolve with the needs of users and now the Echo device can do many other things including controlling household lights and gadgets, allowing users to use the device as an intercom, making household announcements, and playing "whole house audio" for households with multiple Echo devices.

For example, one feature of the evolved Echo skillset is the ability to "drop in" unannounced on the rooms of one's home that have the devices installed. While such a feature is certainly useful when performing common household tasks like announcing dinnertime to young children playing in their rooms upstairs, one can also imagine the moral complications of spying on teenagers and intruding upon the privacy of other adults in the household. Since the modern versions of the device now also allow video, the implications for privacy and surveillance are even more significant and potentially problematic. If others were to gain access to one's Amazon credentials, even complete strangers would be able to "drop in" and observe the family in their everyday rituals or do other problematic things, like impersonate the user and make phone calls on their behalf (Deahl, 2017). The moral dimensions of the Echo example are even more salient when you consider the device's potential interactions with children.

As the Echo example demonstrates, advancements in programmed features and increased usability carry risks to our privacy and security which can be evaluated from an ethical perspective. Webcam hacking and "wardriving" by driving around looking for unsecured Wi-Fi access points, for example, are examples of how our own technologies can be co-opted and turned against us, often for immoral or illegal purposes. The film *Unfriended: Dark Web* (Susco, 2018) plays on these fears and takes this concept to a frightening level. The plot of the film involves the implication of a group of online friends in various murders by an outside person taking over their computing devices and leaving false trails of digital evidence. This is a somewhat farfetched fictional example, but one that also pulls from real fears about these technologies and how programs designed to improve our lives can be coopted for illegitimate purposes. Unfortunately, real world examples of this exist, too; one example is the case in which a baby monitor in Texas was hacked, allowing strangers to both view imagery of a two-year-old child and communicate with her. Disturbingly, the hacker "used the device to curse and say sexually explicit things to the sleeping girl—calling her by name and telling her to wake up" (Ngak, 2013). The programmability and customization of Internet-connected devices such as webcams, baby monitors, and voice-controlled devices, while convenient and helpful for legitimate users, also affords these types of security concerns.

Even more fascinating, perhaps, are programmed algorithms that *change themselves* over time. For example, genetic algorithms and machine learning software use algorithms to learn from mistakes and refine their functionality in order to improve from generation to generation (modern computers process billions of instructions per second, so these generations can occur within microseconds of one another). Machine learning algorithms are useful for features like facial recognition or voice recognition, where time or long-term environmental factors may change the look of one's face or the cadence of one's speech patterns. The artificial aging applications we discussed in this chapter's introduction are also examples of this. Other specialized algorithms may be used in situations where the problem domain is still unknown or uncertain. Genetic algorithms, for example, have been researched as tools for breaking file encryption since decrypting ciphers is an arduous task made even more complicated by the unknown cipher key used during the encryption process. Some researchers have speculated that codebreaking algorithms that can adapt or change over time, by using fitness functions to weed out unsuccessful algorithms and devote additional processing time to those algorithms that show promise, may be our best hopes of breaking complex encryption. And of course, that very act of codebreaking suggests an entirely new set of ethical conversations for us to pursue.

Procedurality and Black Box Computing

When we consider the possibility of moral algorithms, we must review these sequences of instructions not only as static entities, but also as programs that are *active*. In other words, algorithms are not just individual steps that exist in isolation to one another; rather, they are procedures in which each step depends on the prior step for proper operation. Procedurality is a concept closely related to programmability. Procedurality means that when computers are processing instructions, they follow certain steps, steps encapsulated in "units" or "blocks" of code. This traversal of steps is done in an order that is meaningful to the compiler or interpreter so that the appropriate steps are performed in the appropriate sequence (one can imagine the chaos caused if mathematical calculations were computed in the wrong order, for example). These units of code are lumped together into procedures, or associated units of code that are logically connected. Such clusters of code may be called things like procedures, functions, or methods, depending on the programming language being used and how the program is constructed.

Procedures and programmability work hand in hand in that most programming languages have built-in functionality for easily "running" procedures and for doing specialized operations with them, such as repeating instructions (iteration) or only running instructions in certain conditions (conditional evaluation). For example, high-level programming languages such as C, C++, and Java have language constructs known as loops that allow the execution of procedures (or individual statements) repeatedly until certain conditions or met, or while a certain condition is true. These loops might take the following logical form:

1. Set data to an initial value.
2. While that data is below a boundary value, do the procedure.
3. Check to see if the boundary value is exceeded. If it is not, keep going. If it is, stop.

Another common procedural loop's logical format is as follows:

1. If data is below a certain value, do the procedure.
2. Is current data value above the boundary value now? If so, stop. If not, continue.

It is not important that every human developer working on a digital software project understand the inner workings of each procedure she uses. In fact, the abstraction of the inner functionality is seen as a design advantage since there are often many individuals working together on a single project and each developer only need know the appropriate inputs and the expected outputs of each procedure in order to use the algorithm elsewhere in the software. This so-called "black box" model of computational processing means that given a set of valid inputs, the specific code inside the "black box" can be largely ignored if we receive the outputs we expect.

This philosophy of procedural programming means that we can build incredibly complex software systems without needing to be experts on every specialized portion of a software or hardware system. However, there is also a potential ethical complication with this sort of paradigm. When computers execute procedures, there is no central moral authority making judgements before code is processed. In other words, a procedure used to sort through database fields to identify high-income earners could be used for a program intended for fundraising or one intended to flag individuals for potential audits by the Internal Revenue Service. Similarly, a hardware system using a procedure designed to measure the flow rate of gas could be used in a system that fertilizes crops or one that deploys chemical weapons.

Another significant implication of the procedurality of digital media relates to its ability to be copied exactly from one virtual location to another. Ess (2014) discusses this phenomenon and defines it as one of the central issues in the ethics of digital media, noting that in contrast to analogue media (e.g., vinyl records), digital media procedures allow the *exact* copying of content and the *immediately transferrable* and *exactly replicable* transmission of digital content across geographically vast networks. Because of the further compactness allowed by data compression techniques, this means that digital media procedures can transmit significant amounts of content quickly and efficiently. We previously discussed the ethical implications of reproducibility and transferability in Chapter 4. However, it is worth mentioning here the fact that procedures and functions can be easily packaged and transmitted across networks just as easily as the data they manipulate. This presents a multitude of potential security hazards such as the exchange of programs by "script kiddies," or inexperienced users who can still do a lot of damage (Barber, 2001). These users can use compiled programs and procedures to share easy-to-configure programmatic exploits with others over the Internet. On the other hand, the portability of procedures also affords the opportunity to exchange what many would consider to be more virtuous procedures designed for social good. One example of this can be found in the crowdsourced efforts to help do research into protein folding, as found in initiatives such as the Foldit Cookbook, which allows players in a scientific discovery game to organize and disseminate their procedural recipes for gameplay and customization (Cooper et al., 2011).

The Personalized and the Procedural

One way to contemplate the moral dimensions of the programs and procedures of digital technology is to consider the ongoing relationship between ourselves and the technologies we create and use. We, as owners of digital devices, use programmed technologies in nearly all aspects of our lives—as tools for problem solving, communication, organization, and entertainment. If our functional needs in these areas are not met, then the technologies will be abandoned. This means that the ongoing interactions between the many dimensions of human experience (e.g., biological, cognitive, emotional, social, etc.) and our procedural digital technologies are critical to understand. In addition to the humans who program the computers and create the algorithms, we (as end users) also often configure, alter, and extend procedures to personalize and apply these technologies to our own lives. The way one family uses an Echo device does not necessarily relate to the way another household uses it, for example.

Researchers have begun paying more attention to this personalized relationship to the digital as technologies have matured, with topics such as "human–computer interaction," "user experience," and "persuasive computing" all receiving significant attention in the literature. One relevant area of research concerning the human relationship to procedurality and programmability comes from research that studies how videogames influence human thinking and behavior. As developers of highly interactive and immersive technologies, videogame companies have (collectively and over time) invested billions of dollars to determine how to keep humans motivated and engaged using technology. Well-designed games encourage players to continue interacting with their game consoles over prolonged periods, sometimes hundreds of hours or more. When we participate in these prolonged digital interactions, we also become intimately familiar with the procedures and programs that undergird the software systems. The decisions we make as we play the games also shape how future programming unfolds and how future procedures adapt to our decisions. The decisions made by one group of people—the game's designers and developers—can also influence how another group of people—the game's players—think about certain systems (e.g., non-digital procedures and processes that exist in everyday life). In this case, the designers and developers create and implement the algorithms and the players experience and are influenced by them. There is a persuasive dynamic at work here and one way of considering this phenomenon is through the lens of procedural rhetoric.

Ian Bogost, a computer games researcher at the Georgia Institute of Technology, defines procedural rhetoric in his 2007 book, *Persuasive Games,* in several different but related ways, as "a practice of using processes persuasively," as "the practice of persuading through processes in general and computational processes in particular," and as "a technique for making arguments with computational systems and for unpacking computational arguments others have created" (2007, p. 3). He begins his book by discussing an example of an early videogame created by Owen Gaede in 1975 called *Tenure.* The game positions the player as a new high school teacher who is attempting to secure a teaching contract and learn the organizational, political, and social systems at work within a high school teaching environment.

The point of the *Tenure* videogame is to "complete the first year of teaching and earn a contract renewal for the next" (Bogost, 2007, p. 1). In his analysis of the game, Bogost notes that "*Tenure* outlines the *process* by which high schools really run, and it makes a convincing argument that personal politics indelibly mark the learning experience" (p. 2). After playing the game for a sufficient amount of time, a player of *Tenure* who is not already a high school teacher may gain a new understanding of what it is like to work in a high school, or perhaps they may find themselves possessing more empathy for high school teachers they know in the "real world."

Another example of procedural rhetoric is found in a game called *Cart Life,* which is a game about running a street cart as a vendor: managing inventory, dealing with customers, and grappling with all the myriad complications that accompany this type of business. As Fernández-Vara (2015) explains, the game's procedural responses to decisions made in the game make it difficult for the player to be successful if she chooses to adopt the same types of strategies that have proven successful in other types of simulation strategy games. She writes:

> *Cart Life* is a very peculiar artifact – a simulation of running a street cart that turns some of the conventions of strategy games against the player, by not providing the player with help or power-ups, or even an agenda to remember what to do; time passes and it is very challenging to achieve all of the goals set for the day. *Cart Life* is an example of *procedural rhetoric,* where the procedures of the game imply and convey a certain meaning as the player plays the game; it is therefore necessary to explain not only the novel mechanics but also what they mean.
>
> *(Fernández-Vara, 2015, p. 96)*

What Bogost and Fernández-Vara and others who study videogames have found is that these procedural actions can have a significant impact on our understandings of complex systems. By participating in the procedures virtually, we are given new perspectives into complex systems and afforded opportunities to act within those systems in virtual environments with drastically reduced consequences. No real harm will come to us and our physical bodies are not in peril, so we are encouraged to experiment outside our normal parameters. Even more significantly, our in-game actions and decisions become part of the structure of implicit or explicit procedural arguments developed by a game's designers. In other words, we can see the implications of our decisions and actions reflected in a larger system made possible by the videogame's virtual environment. In addition to *Tenure*'s ability to teach us about high school politics, for example, we can also learn from other games about the impact of foreign policy decisions on terrorism or explore the social dynamics of the days leading up to school shootings or tragedies like 9/11. In addition to learning about the mechanics of running a street cart in *Cart Life*, we also learn about the social and political forces that shape life as a street vendor and we may even develop empathy for these business owners that we did not previously have.

Another powerful example of procedural rhetoric is found in a TED talk given by game designer Brenda Brathwaite (2012). Brathwaite is well known in the academic games community for, among other things, creating a game called *Train* (2009) where players removed the game's rules from an authentic Nazi typewriter and then "are told to use toy trains to transport bright yellow tokens from one end of the board to another" (Tsoulis-Reay, 2010, para. 6). At the end of the experience, players are shocked to discover that they have been taking part in the virtual transportation of Jewish prisoners to Auschwitz and many players were moved to tears after discovering their role in a virtual genocide (Baker,

2013). As an artist, Brathwaite's game design "stems from her interest in the power of games to express painful emotions, by implicating players in a system of tragedy" (Tsoulis-Reay, 2010, para. 8). As a *Wired* article's title about *Train* proclaimed, "Playing this board game is agony. And that's the point" (Baker, 2013).

In her TED talk Brathwaite described how the concept for another game, this one about the slave trade, originated. She was speaking with her seven-year-old daughter about the Middle Passage, whereby slaves were transported from Africa to America to be sold. As her daughter explained what she learned about the topic in school, Brathwaite joked that she seemed to be absorbing this incredibly significant and emotionally powerful piece of history a little too casually, considering the Middle Passage as though "some black people went on a cruise." Since Brathwaite's daughter has an African-American father, Brathwaite did not feel as though her daughter had learned about the embedded values and significance of this event and when she asked her mother to play a game, they designed a game about the Middle Passage together. What Brathwaite found was that by designing this game with her daughter, concepts related to the Middle Passage became clear in a way that they were not in other ways (e.g., reading a book about the topic or talking about the topic in school).

Ultimately, after painting tokens to represent humans and building rules that specified how many humans could cohabitate together on a boat and for how long, Brathwaite's daughter came to understand that the Middle Passage was not, in fact, anything close to resembling a cruise. By applying the instructions and following the procedures outlined in the game rules, she learned the operational mechanics of the Middle Passage in a very personal and immediate way, finally recognizing the harsher realities such as families being split apart and individuals dying at sea during the voyage. The ultimate game created by Brathwaite and her daughter is not something most of us would describe as fun. However, the game is so educational and illustrative precisely because of the procedures it employs. It is the interaction between game rules and human gameplay that showcases the power of these procedures in a game-based setting. (It is worth noting that although this game was a physical board game, the same type of procedural relationship is found in digital videogames, except the application of rules is done programmatically by computers.)

Even outside of analog and digital games, procedural rhetoric is useful for helping to understand the ethical nuances of complex digital systems. Because so much of our modern technology is interactive, this means that we must touch buttons, swipe fingers, or otherwise squeeze, shake, turn, look at, or speak to our technologies in order to receive the output we would like to receive. We play an active role in the algorithmic process of computers. Without our participation in the procedures executed by the algorithms, they would never reach the end state of the digital transaction. However, what about the other side of the human–computer interaction: the computer? Are there things we might do as designers to implicitly or explicitly instruct our digital devices to adhere to an ethical code? It turns out that some thinking has already been done on this subject and we turn to the topic of machine ethics next.

Autonomous and Semi-Autonomous Procedures

As we will see later in the chapter, another ethical implication of procedurality is that sometimes computational procedures are designed to entirely replace procedures previously done by humans. This removes the need for the human to be there in the first place, at least according to the profit-driven motives of large corporations. One obvious example can be found in assembly line robots that have gradually replaced human machinists in automobile factories. Since repetition and conditional logic are two modes of operation clearly within the wheelhouse of digital robotics, replacing the person on the assembly line who screwed parts onto a transmission assembly seems like an obvious domain in which automated procedures will thrive. However, beyond the obvious economic implications of a person who has now been replaced by a machine and must seek other employment opportunities, there are also new questions raised about how that machine functions in unforeseen circumstances, particularly when the machine has the capacity to operate with little or no human supervision. In these instances, the machines are described as autonomous or semi-autonomous.

What if such a machine on an assembly line malfunctions and sends the wrong part down the line—is the device sophisticated enough to adapt to the new scenario? What if there is a fire or electrical problem or a forklift crashes into the manufacturing room—does the robot have the same ethical obligations to respond with caution and timeliness that a human operator would have? Clearly, the algorithms reacting to such conditions would need at least a basic moral framework from which to evaluate and respond to the situation. Might a sufficiently mobile robot even function better in this sort of emergency, since the emotional complexities of the human condition are removed from the equation? Or are those emotions necessary and useful in these types of situations?

Although it might seem farfetched, there are documented cases of machines killing humans on assembly lines, such as an instance occurring in 1981 when a 37-year-old motorcycle factory employee was killed by an artificially intelligent robot. The robot "erroneously identified the employee as a threat to its mission and calculated that the most efficient way to eliminate this threat was by pushing him into an adjacent operating machine" (Hallevy, 2010, p. 172; see also Gunkel, 2012). The robot used a powerful hydraulic arm and "smashed the surprised worker into the operating machine, killing him instantly" (p. 172). Upon elimination of the "threat," the robot resumed its duties. According to its programming and the associated normative goals of being as efficient as possible with its manufacturing duties, the robot followed the proper procedures. However, from a moral perspective, the enacted procedure was horrifying. Even a basic moral algorithm, perhaps following principles from the ethics of care (discussed in Chapter 2) in which robots see themselves as relational beings responsible not only for the manufactured product but also the other employees on the assembly line, would have prevented such an accident. It is more than likely that primitive rules for protecting human lives were indeed programmed in, but clearly they were not foolproof enough to account for all variations of human interaction on the line.

While we might characterize the motorcycle assembly robot's killing as a freak accident occurring in the early days of robotics, this is unfortunately not the only example of machines killing humans through procedures run amok. In modern times autonomous machines of war are being designed expressly for this purpose, and sometimes they too make mistakes. Wallach and Allen (2009) write of an instance in October of 2007 when "a semiautonomous robotic cannon deployed by the South African army malfunctioned, killing 9 soldiers and wounding 14 others—although early reports conflicted about whether it was a software or hardware malfunction" (p. 4). Whether a software or hardware issue caused the fault, clearly a procedure somewhere was deficient in protecting human life (a critique that could be made more broadly against even properly functioning war machines, of course). Complicating the issue of autonomous war robots is the argument made by some that such robots can in fact be *more humane* than human soldiers precisely because they are rule-bound to follow the rules of war, such as limiting noncombat casualties and avoiding the need for "self-preservation as a foremost drive" (Arkin, 2009, p. 31). Additionally, robots have better sensory capabilities, expanding their ability to process complex environmental information, and they can be "designed without emotions that cloud their judgement or result in anger and frustration with ongoing battlefield events." (Arkin, 2009, p. 31). Such commentary reminds us that a series of steps completed by a human and a series of steps completed by a machine must be planned for and evaluated differently because of both these strengths and deficiencies of digital devices.

Life and death issues aside, there are many associated ethical threads we can also follow from considering robots with autonomous or semi-autonomous capabilities. For example, in occupational roles, there are ancillary ethical questions, such as whether robotic workers are trained (programmed) to surveil the humans who are also working in that environment (and whether the human employees are aware of this surveillance). These types of functions point to the existence of *concealed procedures* that may not be visible to human observers, but that are still occurring nonetheless. These secretive operations have direct implications for our agency and autonomy; as Moor (2006) notes, ethical issues dealing with privacy, property, and power are all significant when we design computers to "do our bidding as surrogate agents" (p. 19). We will engage a bit more with Moor's ideas about machine logic later in the chapter, but before we do that, we will first examine another aspect of algorithms, which is the potential role of computational procedures in our thinking and reasoning in cooperative environments where humans and machines work together for a common purpose.

Cooperative Computing and Machine Ethics

As we questioned in the previous section, given the importance of software programs to our digital technologies, might we develop machines to act in morally agreeable ways? Is there a way to develop technologies that use ethical algorithms and procedures to make decisions? Some research has explored these questions using specific ethical paradigms. For example, in *Ethics for Robots: How to Design a Moral Algorithm,* Leben (2019) suggests evaluating algorithms by their cooperative abilities, as evaluated through contractarianism (we previously discussed contractarian ethics in Chapter 2). Ultimately, Leben argues that "our moral intuitions are the product of a psychological network that adapted in response to the problem of enforcing cooperative behavior among self-interested organisms" (2018, p. 5). This focus on cooperation through contrarian ethics is well-suited for computers and machines; as we noted in Chapter 2, since morality is linked to an agreement of basic principles for guiding social interaction, then a goal-directed scenario for evaluating whether such principles are being followed makes good sense.

Ultimately, Leben proposes the design of an "ethics engine" based on contractarianism using chess software programs as a model. In comparing an ethics engine to a chess engine, Leben notes the algorithmic similarities. Each involves a process in which the program must "represent all the current objects within the current state, map out possible actions and outcomes, assign a value to each outcome, and then use a rule for picking which action is best" (p. 77). As Leben also notes, "Each of these steps involves important moral assumptions" (p. 77). For example, moving back into our artificial aging applications discussed at the beginning of the chapter, each of these actions can be considered from an ethical perspective. Representing all objects within the current state may involve facial detection algorithms that we have already shown can have bias. Mapping out possible actions and outcomes may mean evaluating different graphical representations of aging and making decisions based on incomplete data extrapolated from existing physical cues. Assigning a value and picking which action is best means making decisions along both explicit (e.g., which algorithmic visualization is likely to be more aesthetically pleasing?) and implicit (e.g., where does the biometric data go once it is collected?) avenues. The evaluation of such algorithms according to a goal-directed cooperative rubric is also interesting; while a very user-friendly and accessible product can be evaluated as highly cooperative—and morally sound—on a surface level, if data is being collected and distributed without consent beneath the surface, then these algorithms are failing to meet the expectations for moral behavior in another area of the software's operation. It is these lower-level software transactions, occurring beneath the visible layer of interface and surface functionality, that are often the most ethically problematic in digital software.

Machine ethics is another field that lends insight into moral algorithms. In contrast to issues in *computer ethics,* which involve such topics as computer hacking, software property, and privacy, *machine ethics* focuses on how machines behave toward humans and other machines (Anderson & Anderson, 2007, p. 1). Stated another way, machine ethics is about "developing ethics for machines, in contrast to developing ethics for human beings who use machines" (Anderson & Anderson, 2011, p. 1) and machine morality "encompasses questions about what moral capacities a robot should have and how these capacities should be computationally implemented" (Malle, 2016, p. 243). This research necessarily brings in ideas from both philosophy and computer science in order to consider questions of machine ethics. For example, there are a multitude of algorithmic machine ethics implementations based around specific objectives, such as discerning truth telling abilities from software presented with ethical dilemmas or following specific ethical paradigms. An example is the JEREMY system, which makes utilitarian decisions under the assumption that action is morally right if it maximizes good consequences and "the pleasure and displeasure of those affected by each possible action are considered" (Pereira & Saptawijaya, 2016, pp. 8–9). When reviewing their examples, Pereria and Saptawijaya (2016) organize them along two branches of machine ethics research—one realm in which computation itself is a "vehicle for representing moral cognition of an agent and its reasoning thereof" (p. 7) and another where stable moral norms evolve over time within a given population of software agents. The distinction is important; the former approach results in algorithms in which ethical paradigms are built into the initial codebase, while the latter allows for an evolutionary approach using techniques like machine learning to adjust algorithms over time to accommodate changing trends within a population or environment.

Machine ethics is of critical importance for autonomous devices in particular; scenarios in which machines do work without human intervention require us to trust the machines enough to make good (and safe) decisions and enact proper procedures on their own without real-time input from human operators. An adequate model of machine ethics in the motorcycle assembly line robot in 1981 would likely have prevented death for the assembly line worker.

Closely related to procedures of reasoning are procedures of agency. If a machine can reason through an ethical dilemma, is it empowered to act upon that reasoning to prevent tragedy from occurring? Indeed, a central consideration of machine ethics is the question of machine moral agency and this consideration also raises questions about who is responsible when disasters occur. As Gunkel (2012) asks, does the location and assignment of responsibility lie with a human developer, or with the machine itself? Although HAL from *2001: A Space Odyssey* (Kubrick, 1968) is the stereotypical example from science fiction of machine deception, Gunkel (2012) recounts several contemporary—and real, non-fictional—examples pointing to a need for morally responsible machines, including the assembly line employee and autonomous cannons we previously discussed.

We can additionally consider the relationship between machine ethics and other types of more abstract procedures, such as the cultural activities that surround formerly analog undertakings that are now mediated by digital technologies. Consider the work of James Moor, a faculty member at Dartmouth College and another longtime researcher of machine ethics. Moor (2006) reminds us that we can evaluate and assess digital technologies not only according to design norms, but also according to ethical norms. He uses the example of robotic jockeys for camel racing that were developed to replace young Sudanese boys who were mistreated or underfed (Lewis, 2005). Moor notes that the technology was developed in Qatar, where camel racing is a popular pastime. The robotic jockeys are sophisticated. They "are about two feet high and weigh 35 pounds. The robotic jockey's right hand handles the whip, and its left handles the reigns. It runs Linux, communicates at 2.4 GHz, and has a GPS-enabled camel heart-rate monitor" (Moor, 2006, p. 19). Evaluated according to design norms, the robotic jockeys were successful, in that they successfully performed the same duties as a human jockey. Evaluated according to ethical norms, they were also successful in that each robotic camel jockey helped free a young Sudanese boy from slavery. However, as Moor notes, "although this eliminates the camel jockey slave problem in Qatar, it doesn't improve the economic and social conditions in places like Sudan" (p. 19). While the immediate processes were well-served by this replacement, the broader cultural processes surrounding the process were still problematic. Although they were not entirely successful at removing the connected problems we might identify in this example, the robotic camel jockeys were examples of what Moor calls *ethical impact agents*, or technologies that can be evaluated according to ethical norms.

Other procedural implications of machine ethics relate to more intimate issues of personal security. In his work Moor further defines *implicit ethical agents* as machines that work in safety or reliability settings and must deal implicitly with the ethical implications of those settings. He uses the example of an ATM machine as an implicit ethical agent. Because an ATM is essentially an agent for a bank teller, and because money itself is loaded with ethical implications, the transactions between a customer withdrawing or depositing money and the machine acting as a virtual teller are ethically significant. However, the programming code that is enabling the machine to function is not programmed in an ethically explicit way. Rather, the machine is designed to accept or dispense money accurately and with good feedback given to the customer, helping to assure them of their privacy and putting them at ease that their money is being appropriately cared for by the bank. Similarly, a machine like an airplane autopilot system is an implicit ethical agent because it must be designed to deliver passengers safely and on time or provide appropriate warnings to the pilots if safety hazards are encountered (Moor, 2006).

Most of the digital machines we are familiar with in modern society are implicit ethical agents, if we evaluate their programming or procedural functionality using moral parameters. However, Moor also speculates about *explicit ethical agents*, which would be machines that apply ethical models explicitly, with their built-in programming fashioned to consider ethical models during decision making. He cites the work of two researchers, Jeroen van der Hoven and Gert-Jan Lockhorst, who developed different types of logic that can be used by machines to consider ethical models. They included logics such as a *deontic* logic "for statements of permission and obligation," an *epistemic* logic, "for statements of belief and knowledge," and an *action* logic, for "statements about actions" (2006, p. 20). (You can find the theoretical inspirations for these particular logics by revisiting the ethical models we considered in Chapter 2). In some of our modern cell phone applications built for mobile devices, we are starting to see some of these logics at work. For example, many modern applications are now required to ask the user for specific permissions (e.g., access to one's camera, or access to one's user profile on social media) before the operating system will allow the program to be installed.

The ultimate question in machine ethics, of course, is whether a machine might ultimately be capable of becoming a *full ethical agent,* or of having moral reasoning capabilities similar to a human being (Moor, 2006). This subject has long been debated in artificial intelligence communities, and many believe that since our computers lack intentionality, consciousness, and free will, that such an outcome is highly unlikely, if not entirely impossible. However, as Moor notes, "we can't say with certainty that future machines will lack these features" (pp. 20–21). We can imagine, for example, future programmable machines that are much more advanced than our current computers and that may be capable of encountering and acting within scenarios that are much more ethically signifi-cant than what they encounter in their current domains. Moor notes that under-standing machine ethics and developing ethical agents is not only important because we want our machines to treat us well and we need to properly guide their increasingly sophisticated and powerful functionality, but also because pro-gramming machines to act ethically will help improve our understanding of eth-ics in general. After all, to build an algorithm to properly solve a problem, you need to understand the problem in sufficient detail.

Advanced Robotics and the Ethics of Digital Companions

Speculating about cooperative computing, moral algorithms, and machine ethics naturally leads us down the path of considering new programs and procedures for advanced robotics. How will the robots of the future evolve in their algorith-mic approach to ethics? One would hope that moral algorithms would evolve in parallel with advancements in hardware; after all, we do not want machines—machines that are more powerful than us in many ways—to be murderous, self-ish, or reckless. In a 1942 short story titled "Runaround," science fiction writer Isaac Asimov famously formulated the three laws of robotics, which are as fol-lows (Deng, 2015):

1. A robot may not injure a human being or, through inaction, allow a human being to come to harm;
2. A robot must obey the orders given it by human beings, except where such orders would conflict with the First Law; and
3. A robot must protect its own existence as long as such protection does not conflict with the First or Second Laws.

We have seen the results of when these laws are violated, as shown in the exam-ple of the motorcycle employee killed by a robot who deemed him a threat to efficient operations. In other cases, such as the autonomous vehicle examples we discussed earlier in the book, the ethical imperative of a robot to protect human life is somewhat more complicated because there are human lives both inside and outside of the vehicle, all of which must be accounted for. Examples such as these show that the mandate to protect human life or minimize harm are often quite complicated in non-hypothetical scenarios.

Ultimately, while Asimov's rules are useful for thinking about broad parameters for a robot's operation, he was a science fiction writer and a storyteller, not an engineer or computer scientist. Wallach and Allen (2009) pursue a detailed investigation of Asimov's laws as applied to ethical robot development, ultimately concluding that "they offer little practical guidance as moral philosophy, and their value as specifications for algorithms is questionable" (p. 104). This position is congruent with Anderson (2011) who notes that, among other problems with the laws, intelligent machines "have an advantage over human beings in having the potential to be ideal ethical agents, because human beings' actions are often driven by irrational emotions" (p. 286). Wallach and Allen (2009) further note the importance of such differences, for example in appreciating the utility of Asimov's three laws in pointing out that the behavior of robots "should conform to different standards than the usual rules of morality for human beings" (p. 104). As a procedural template for moral behavior, however, Asimov's laws of robotics are dramatically interesting, but fall short of the necessary safeguards needed for ethical autonomous machines.

Other types of advanced robots are starting to become viable in contexts such as healthcare and the military, places in which human life and safety can be directly affected through the application and evaluation of very fast algorithms and procedures. At some point within their deployment, these robots will likely have to make decisions between two or more bad choices rather than between a clearly good choice and a clearly bad one. When a robotic vacuum cleaner encounters an unknown object in its path it can steer around it or simply stop and wait for human intervention. When a medical robot encounters an unknown pathogen in a medical emergency, stopping and waiting for a human to respond may be a life-threatening decision.

Social robots also pose ethical challenges to algorithm design, particularly when such robots take on the role of caregiver or companion. Consider the case of Nao, a commercial toy robot manufactured by a company named SoftBank Robotics and described as "our first humanoid robot" who is designed to be "an interactive companion robot" (SoftBank, 2017). Because this robot is explicitly designed for human companionship, ethical conundrums can arise quite easily in procedures encapsulating its everyday interactions with human beings. For example, Deng (2015) cites the work of interdisciplinary researchers who bridged research from philosophy and computer science to explore the ethical implications of programmed procedures within Nao. These researchers programmed Nao to remind their owners to take medicine. However, they soon found interesting ethical dilemmas even within that limited context of interaction. In their own words:

> "On the face of it, this sounds simple," says Susan Leigh Anderson, a philosopher at the University of Connecticut in Stamford who did the work with her husband, computer scientist Michael Anderson of the University of Hartford in Connecticut. "But even in this kind of limited task, there are nontrivial ethics questions involved: For example, how should Nao proceed if a patient refuses her medication? Allowing her to skip a dose could cause harm. But insisting that she take it would impinge on her autonomy."
>
> *(Deng, 2015, para. 8)*

Researchers in California have recently outlined a series of categories in which robots like Nao are likely to encounter ethical issues (Lin, Abney, & Bekey, 2011). There are significant challenges robots will encounter in areas including safety, for example, which will require us to carefully consider what is meant by "harm" and how broad this term can be interpreted (think of the economic harm caused by a machinist's replacement by an assembly line robot). Other challenging areas will be in determining rights and responsibilities of robots, how they fit into legal structures and deal with privacy issues, and how they more broadly influence society and culture.

In addition to more trivial functions such as serving as a household assistant or butler, a companion robot could also theoretically develop into something much more. This theme has been explored extensively in science fiction. One example is found in an episode of the Twilight Zone, "The Lonely" (Serling & Smight, 1959), where a man sentenced to solitary confinement on an asteroid is given a robotic companion and thinks of "her" as a human and has an emotional connection. At the end of the episode, a group of astronauts come to the asteroid to take him home and, because there is no room for her on the rocket, one of the astronauts shoots her, exposing wires and metal parts. It is not clear at the end of the episode whether the prisoner continues to think of the robot as a person. Another is the more recent film *Her* (Jonze et al., 2014) where a man has a relationship with an AI and the complications you might expect in this sort of relationship do indeed develop.

As these real and fictional robotics examples remind us, companionship is complex and may mean different things to different individuals who have different expectations and emotional, physical, or cognitive needs in their relationships and friendships. An advanced companion robot could serve as a surrogate for or extension to existing human relationships and it is not unreasonable to think deeper feelings of friendship or even love might ultimately develop between a human and a long-term robot companion. Even physical needs and/or sexual relations could manifest in that relationship, depending on the functionality of the robot. If such devices emerge, they will give new perspective to the types of programs and procedures we assign to our robots as well as open entirely new possibilities for moral case studies.

CASE STUDY: MY BEST FRIEND IS A ROBOT

Considering the long-term possibilities of companionship robots and what they might be asked to do for their owners, can you identify several examples of situations in which a robot might be forced to rely upon programming that directs it to choose between several imperfect procedures in order to deal with a human being? What ethical issues from this chapter or elsewhere in the book are at stake in these types of programmed behaviors? What types of procedures would be important to include in such a robot, and which types of procedures should be carefully analyzed, from an ethical perspective, for the purposes of meeting a wide array of different expectations about companionship? Are there ethical design tactics that could partially address, resolve, or improve such decisions for future interactions?

Next Up

In the next chapter, we discuss another property of digital technology, *embeddedness*, and its own ethical questions for the digital. We explore the implications of digital information and its technologies embedded in our lives and experiences and consider how this intermingling of humans and technologies creates ethically significant moments and interactions.

References

Anderson, S. L. (2011). The unacceptability of Asimov's three laws of robotics as a basis for machine ethics. In M. Anderson and S. L. Anderson (Eds.), *Machine ethics* (pp. 285–296). Cambridge, UK: Cambridge University Press.

Anderson, M., & Anderson, S. L. (2007). The status of machine ethics: a report from the AAAI Symposium. *Minds and Machines, 17*(1), 1–10.

Anderson, M., & Anderson, S. L. (Eds.). (2011). *Machine ethics*. Cambridge, UK: Cambridge University Press.

Arkin, R. C. (2009). Ethical robots in warfare. *IEEE Technology and Society Magazine, 28*(1), 30–33.

Barber, R. (2001). Hackers profiled—who are they and what are their motivations? *Computer Fraud & Security, 2001*(2), 14–17.

Baker, C. (2013, December 23). Playing this board game is agony. And that's the point. *Wired.* Retrieved from https://www.wired.com/2013/12/brenda-romero/.

Bogost, I. (2007). *Persuasive games: The expressive power of videogames.* Cambridge: The MIT Press.

Brathwaite, B. (2012, February). *Games for a change.* TEDxPhoenix. Retrieved from https://www.youtube.com/watch?v=y9Z-3mz3j6U.

Buolamwini, J. (2016, November). *How I'm fighting bias in algorithms.* TEDxBeaconStreet. Retrieved from https://www.ted.com/talks/joy_buolamwini_how:i_m_fighting_bias_in_algorithms#t-15174.

Chesney, R., & Citron, D. K. (2018). Deep fakes: a looming challenge for privacy, democracy, and national security (July 14, 2018). 107 California Law Review (2019, Forthcoming); U of Texas Law, Public Law Research Paper No. 692; U of Maryland Legal Studies Research Paper No. 2018-21. Retrieved from https://ssrn.com/abstract=3213954 or http://dx.doi.org/10.2139/ssrn.3213954.

Cooper, S., Khatib, F., Makedon, I., Lu, H., Barbero, J., Baker, D., ... Popović, Z. (2011, June). Analysis of social gameplay macros in the Foldit cookbook. In M. Cavazza (Ed.), *Proceedings of the 6th international conference on foundations of digital games*, Bordeaux, France (pp. 9–14). ACM.

Crawford, K. (2016). Artificial intelligence's white guy problem. *The New York Times,* pp. 25. Retrieved from https://www.nytimes.com/2016/06/26/opinion/sunday/artificial-intelligences-white-guy-problem.html.

Deahl, D. (2017, June 26). Alexa's new calling feature means it's really time to set up two-factor authentication. *The Verge.* Retrieved from https://www.theverge.com/2017/6/26/15874932/amazon-echo-alexa-calling-security-two-factor-authentication.

Deng, B. (2015, July 1). Machine ethics: The robot's dilemma. *Nature.* Retrieved from https://www.nature.com/news/machine-ethics-the-robot-s-dilemma-1.17881.

Ess, C. (2014). *Digital media ethics.* Cambridge: Polity.

Fernández-Vara, C. (2015). *Introduction to game analysis.* New York: Routledge.

Gunkel, D. J. (2012). *The machine question: Critical perspectives on AI, robots, and ethics.* Cambridge, MA: MIT Press.

Hallevy, G. (2010). The criminal liability of artificial intelligence entities-from science fiction to legal social control. *Akron Intellectual Property Journal, 4*(2), 171–219.

Jonze, S. (Director, Producer, & Writer), Ellison, M., & Landay, V. (Producers) (2014). *Her* [Motion Picture]. United States: Warner Home Video.

Kubrick, S. (Director, Producer, & Writer), & Clarke, A. C. (Writer) (1968). *2001: A Space Odyssey* [Motion Picture]. United States: Metro-Goldwyn-Mayer.

Leben, D. (2019). *Ethics for robots: How to design a moral algorithm.* Oxford, UK: Routledge.

Lewis, J. (2005). Robots of Arabia: The ideal camel jockey is the size and weight of a starving 4-year-old boy. Tradition collides with technology atop a beast racing at 25 miles per hour. *Wired, 13*(11), 188.

Lin, P., Abney, K., & Bekey, G. (2011). Robot ethics: Mapping the issues for a mechanized world. *Artificial Intelligence, 175*(56), 942–949.

Malle, B. F. (2016). Integrating robot ethics and machine morality: the study and design of moral competence in robots. *Ethics and Information Technology, 18*(4), 243–256.

Moor, J. H. (2006). The nature, importance, and difficulty of machine ethics. *IEEE intelligent systems, 21*(4), 18–21.

Moschovakis, Y. N. (2001). What is an algorithm? In *Mathematics unlimited—2001 and beyond* (pp. 919–936). Springer Berlin Heidelberg.

Ngak, C. (2013, August 13). *CBS News.* Baby monitor hacked, spies on Texas child. Retrieved from https://www.cbsnews.com/news/baby-monitor-hacked-spies-on-texas-child/.

Noble, S. U. (2018). *Algorithms of oppression: How search engines reinforce racism.* New York, NY: NYU Press. [Kindle edition].

Pereira, L. M., & Saptawijaya, A. (2016). *Programming machine ethics.* Switzerland: Springer.

Serling, R. (Writer) & Smight, J. (Director). (1959). The lonely. [*Television series episode*]. In B. Houghton (Producer) *The twilight zone.* Los Angeles, CA: Cayuga productions and CBS Television Network.

SoftBank Robotics (2017). Who is Nao? Retrieved from https://www.ald.softbankrobotics.com/en/robots/nao.

Susco, S. (Writer & Director). (2018). *Unfriended: Dark web.* [Motion Picture]. Universal Studios.

Thompson, A.C. (2018). In Amazon Go, no one thinks I'm stealing. *C/Net.* Retrieved from https://www.cnet.com/news/amazon-go-avoid-discrimination-shopping-commentary/.

Tsoulis-Reay, A. (2010, May 25). Brenda Brathwaite: Message in the machine. *Popmatters.* Retrieved from https://www.popmatters.com/123880-brenda-brathwaite-message-in-the-machine-2496195309.html.

Tucker, I. (2017, May 28). 'A white mask worked better': Why algorithms are not colour blind. *The Guardian.* Retrieved from https://www.theguardian.com/technology/2017/may/28/joy-buolamwini-when-algorithms-are-racist-facial-recognition-bias.

Wachter-Boettcher, S. (2017). *Technically wrong: Sexist apps, biased algorithms, and other threats of toxic tech.* New York: WW Norton & Company.

Wallach, W., & Allen, C. (2009). *Moral machines: Teaching robots right from wrong.* Oxford, UK: Oxford University Press.

6

THE DIGITAL AND THE HUMAN (EMBEDDEDNESS)

In this chapter we explore the ethical implications of embeddedness in our relationships with technology and discuss how embeddedness influences our decisions and behaviors in digital spaces. Such discussions raise questions about the nature and function of identity, privacy, and anonymity in relation to digital technology. They also suggest new possibilities for thinking about our relationships with one another and with technology. The types of technologies we use are embedded with the values of the organizations that design them, indirectly influencing broader cultural and societal practices through their widespread adoption. Consider the fanciful case of Bill the time traveler to stimulate your thinking about the embeddedness of the digital.

CASE STUDY: BILL THE TIME TRAVELER

Imagine Bill, who is a seasoned time traveler. Over the course of the last few years of his life, Bill has met 16th century Chinese emperors, conversed with 18th century British philosophers, shared meals with World War I soldiers, and asked Egyptian pharaohs about the nature of the afterlife. Then one afternoon, thanks to a miscalculation, Bill finds himself in the year 2119, having traveled forward rather than backward in time. Concerns about the physics of time notwithstanding, Bill is now in the future, but at least it is a future with bars. Bill seats himself at a bar and strikes up a conversation with Ted, the person sitting next to him. After talking for several hours about philosophy, life, and why the 2016 U.S. presidential election was a complete train-wreck, Bill realizes that he "gets" Ted: He feels connected, like he understands where he's coming from. Feeling emboldened, Bill shares with Ted that he's a time traveler. Ted nods his head and shares something too: He is, in fact, a machine. Bill felt connected to Ted, but now feels betrayed and a little weirded out.

The hypothetical experience of Ted's feeling of betrayal should make us think of the sorts of ethical questions we addressed in the last chapter on machine ethics and the procedurality of the digital. But another part of the reason for Ted's unease, perhaps, is the ways in which Bill's disclosure makes visible the embeddedness of the digital in everyday experience. Cases like Bill that push our imaginations can be powerful, but we need not stretch very far to get to some of the same ethical issues. Recall again the case of Conrad Roy with which we began Chapter 2. The attorney for defendant Michelle Carter noted, as part of her defense, that Carter was miles away from Mr. Roy when he committed suicide, so she could not be held responsible for his death, much less be considered the cause of it. However, Mr. Roy's sister noted (rather importantly) that she was *as good as there with him* because the use of digital technologies made her voice and what she wrote a part of what it meant to "be there" with Roy.

While the suicide of Mr. Roy is a sad case, not all cases of embeddedness are so dark. In fact, the opposite may be true. Online interactions right now allow opportunities for people with similar interests to connect positively. They may then experience budding friendships at first that grow into the sort of friendship that philosophers like Aristotle, Montaigne, de Beauvoir, and others (who thought about things like friendship quite a bit) considered the most valuable of all human relationships. Furthermore, the embeddedness of our personal lives with each other may very well manifest itself in other positive ways. We might imagine a case like that of Mr. Roy in which someone showing an interest in and care for a young man or woman contemplating suicide would lead him or her to *not* go through with it. Digital technologies and the digital cultures they enable make possible the creation of meaningful relationships—both positive and negative—with each other in at least some contexts that would not have been possible before the emergence and use of digital technologies. They do so precisely because of the ways that they have become embedded throughout our social fabric and most intimate interpersonal spaces.

While Mr. Roy's suicide is a dramatic example of this digital embeddedness, our everyday relationships are also impacted. Many internet users' relationships are created and sustained in online communities, perhaps with the individuals never having met in physical space. Similarly, our real relationships are mediated with and through the devices embedded in our daily routines. It is not unusual for friends or family members to exchange text messages, serendipitous photographs, audio messages, or videos with one another during the course of their daily lives to bring those close to them in on the parts of their lives these people would not normally see. Even the three of us, as we write this book, communicate with each other largely using online productivity and file sharing sites such as Google Docs and Dropbox. We also use the messaging function in Google Docs and Skype not only to talk about the progress and sections of this book, but to share information about unrelated issue such as an interesting new book to read or what we had for dinner. The technologies and orientations to the world that make these digital interactions possible are so deeply embedded in our lived experience that we have to work to draw attention to them.

An Embedded Existence

By "embeddedness" we refer to the ways in which (1) aspects of ourselves, including our behaviors, are integrated into digital information networks, (2) digital technologies are embedded into our personal and professional lived experiences, and (3) institutional values are propagated through the technologies organizations create and disseminate. Each of these perspectives on embeddedness offers interesting insights and potential case studies for ethical reasoning in digital environments.

Let us first consider how embeddedness and ubiquity, or the state of "being everywhere," are related. We sometimes hear the phrase "ubiquitous computing" used to describe the mass utilization and availability of computational technologies throughout society. Digital devices and digital content are indeed almost everywhere in the industrialized parts of the world. They are deeply embedded into our everyday activities and intimately linked with our everyday lives. We increasingly depend on digital technologies for our commercial interests—like banking and shopping—as well as for our daily communication with friends, families, and business associates. Increasingly, we also see digital technologies embedded into our leisure and entertainment activities, even in spaces previously not associated with such practices. For example, people are often seen at concerts and other sorts of public or social events looking down at their phones, sending messages to their friends and family members or looking at social media, essentially ignoring parts—if not all—of the events they are attending.

While we might initially be tempted to criticize such attention-stealing embedded activities, perhaps by commenting on the sad state of affairs for the individuals who are not being fully "present" at such events, digital embeddedness is not something we can immediately classify as immoral or even ethically problematic. Sometimes the attention given to digital devices is part of a quest to bring informational exposure to public or social events or occurrences. For example, it has become commonplace for bystanders to take pictures or videos of police activity and then to post them on social media or deliver them to the news media. When such digital media captures a problematic situation and distributes it to the community in a way that garners a great deal of attention, it may lead to changes in policies or procedures that are more just or equitable for certain individuals or groups.

Embedded digital technologies also enhance the possibilities for conversation and interaction between different audiences and users. Often, these technologies merge different modes of media—such as audio, text, and video—to create dynamic possibilities for teamwork and communication. Some gamers find themselves involved in discussions on the side in apps like Discord and TeamSpeak where what the users of these communication apps are talking about may have nothing to do with the game they are playing. Friendships can be created in online communities of gamers, in online social media sites where people with similar interests find themselves in groups, and in online discussion groups in which interests are shared between members and users of the sites. Family members often create private groups in social media that are specific to family members, and it is possible in social media to restrict access to view files, pictures, posts, and other content.

In each of these examples, embeddedness plays a role in how we understand and respond to such situations in our own lives. For instance, things that seem natural due to their ubiquity, proximity, and familiarity—as embedded technologies increasingly are—are treated differently from things that are alien, unusual, and unknown. We may overlook negative aspects of embedded communication technologies, like group texts and online news articles, because we have integrated them so significantly into our communication processes and our consumption of information about the world. We may not think of or worry as much about the security and privacy of our personal information and demographics due to the convenience of embedded passwords, profiles, and social media systems. And we may become emotionally attached to technologies with which we spend significant amounts of time, as though we have transferred a portion of ourselves into the devices. Embeddedness enables us to be present (see Floridi, 2014) in multiple diverse contexts at the same time, spreading out our capacities as well as our vulnerabilities in dramatic ways.

Selves and Identities

The relationship between identity and technology has long been a target of academic and cultural study. Sherry Turkle's book *The Second Self* (1984/2005) laid the foundation for thinking about computing from a humanist perspective, allowing for explorations of identity, feelings, and morality in examining the interactions between humans and computers. Turkle explored computing culture of the late 1970s and early 1980s by interviewing a broad sample of computer users ranging from children through college students to professional scientists and engineers. She found through these interviews that many of these devices had a more meaningful impact on their users' lives than one might imagine. In contrast to a computer being merely a cold, logical device of mathematics and computing, Turkle argued that the device also possessed a "second nature" as an "evocative object, an object that fascinates, disturbs equanimity, and precipitates thought" (loc. 282–283). As evocative objects, Turkle wrote of the "holding power" of computers, their ability to captivate and immerse users and keep them focused on the technology for long periods of time. And she writes of the "projection of self onto computational media" as a "second self" which allows one to be a different person when represented through a digital computer. The number of avatars, profiles, game characters, Internet identities, and now three-dimensional virtual bodies that vary significantly from real world biological, emotional, and sexual selves are testament to Turkle's prescient analysis more than two decades ago.

This exploration between real and virtual identity has been the subject of much research investigating identity in relation to different types of digital technology. One well known example analyzing the identities formed by video game players is found in James Paul Gee's (2007) book *What Video Games Have to Teach Us about Learning and Literacy*. Gee distinguishes between three different types of identity videogamers experience: a real-world identity, made up of one's own identity as a nonvirtual person; a virtual identity, made up of one's virtual character in the videogame; and a projective identity, made up of the relationship between the two. For example, a videogame player named Jane Doe playing a character named Aloy in the game *Horizon Zero Dawn* has both a real identity as Jane Doe, a virtual identity as Aloy, and a projective identity that is Jane Doe *as* Aloy. Jane Doe's embedded virtual identity as Aloy can be understood in two ways: as a projective relationship transferring one's values onto the virtual character, and as a means of a "project in the making" (p. 50) that can be built up in a certain way to behave in a certain way according to the player's own desires and the affordances built into the game by the game designers. It is interesting to note that the second "project in the making" scenario allows a greater degree of flexibility in the construction of digital identity; one may consider herself to be quite scrupulous in the real world but then choose to imbue her virtual avatar with an immoral conscience and a deep desire for destruction and malfeasance. Many games, such as *Fable* and the *Fallout* series, are highly regarded in part because they allow their players to choose whether they will play as "good" or "evil" virtual characters.

While the rich fantasy and deep interactivity found in videogames allows for compelling transformations of identity in digital space, the complex relationships between our selves, our identities, and our devices are not only occurring in games. For many, these experiments with digital identities began at an early age, as we formed our earliest relationships with technologies. The relationship between identity and technology in young people is well documented and has evolved over different generations of both people and technology. Sometimes, these relationships can be used to differentiate certain demographics from others due to the different technologies with which they grew up and the different ways these technologies were embedded into their lives. For example, cultural anthropologist Mizuko "Mimi" Ito and colleagues (2010) note that the link between generational identity and technological identity is particularly strong in contemporary youth, writing that "there is a growing public discourse (both hopeful and fearful) declaring that young people's use of digital media and communication technologies defines a generational identity distinct from that of their elders" (p. 2). By situating this generational identity within the context of social and cultural studies, Ito and her research colleagues studied the implications of rapid technological change on underlying social and cognitive practices such as sociability, learning, play, and communication.

Ito et al. (2010) draw from Lessig's (2004) work to explore the notion of an ethic of civic responsibility in which communities band together to produce "public rather than proprietary goods" (Ito et al., 2010, p. 325). Given that the youth of today are often networked to one another extensively through social technologies such as Facebook, Twitter, Instagram, and Reddit and videogames like *Minecraft* or *Fortnite*, it is not surprising that this embeddedness has propagated value systems that are shared by many users of similar ages. Complex psychological feelings such as self-efficacy and complex social practices such as teamwork may also be bound up in these early experiences with digital identities. Regarding self-efficacy, playing as a character with different characteristics can increase one's confidence and willingness to take risks, whether those risks are engaging in conversation with other users or tackling a project that seems outside of one's current skill set.

A child struggling to make friends in the third-grade classroom may suddenly be in demand as a squad mate due to a virtual identity establishing his or her videogame prowess in a virtual world. In teamwork, sometimes the reputation component of identities are reinforced in different ways—through endorsement or association with other users' identities from one's guild, gaming group, or forum channel. One can imagine children growing up playing hundreds of hours of games like *Minecraft* or *Fortnight* having positive attitudes towards Lessig and Ito's concept of "public rather than proprietary goods"—things like sharing content with other users, engineering complex systems in teams and making them open source, or sharing resources within a group. However, we more frequently hear about dramatic cases in the news media that position videogames as societal ills that lead to more harm than good. Many of these cases portray videogames as problematic technologies that encourage aggressive or even violent behaviors in their players (e.g., Snider, 2018). It is worth pointing out that such studies have been challenged by other researchers working in the field who found no significant connection between youth aggression, crime, psychopathy, and videogame playing (Smith, Ferguson, & Beaver, 2018; Ferguson, 2015).

It is not only youth, however, that embed portions of their identities into digital technology, and it is not only through playing videogames that this process occurs. Older adults (who also play video games, by the way) are equally capable of trusting their devices with personal information and digital data describing their identities. Consider the amount of personal information teenagers and adults alike now trust to their mobile devices. Not only do these technologies contain detailed lists of our personal and professional contacts, but they also store our musical and film preferences, our social media profiles, and even our geographical histories: where we travel, how long we stay, and when we visit our favorite restaurants and businesses. With newer devices, even our biometric information is saved—our fingerprints, facial profiles, and medical records. In only a short amount of time, perhaps a decade, we as a society have seemingly broadly accepted the practice of transferring large portions of the personal data describing our lives to digital databases.

The ethical implications of the trust we place in our embedded digital lifestyles are significant. Terms of Service (TOS) documents are legal contracts that describe in detail how technologies can be used and how one's data is treated inside those technologies and their associated databases. One could argue that you deserve what you get if you are unhappy with how your data is being used, since it is spelled out for you clearly in those TOS agreements you were required to "accept" before installing the technologies on your computers. However, while it is possible that some users carefully read every TOS document provided by large companies like Apple, Google, and Amazon before using their products, the reality is that the majority do not. The devices and software are too immediately useful to worry much about longer-term issues like privacy, data, and surveillance. Further, many TOS documents are overly long, detailed, and cumbersome; most users just want to "get to it," so to speak. Many are also complicated by excessive amounts of legal jargon and other language everyday computer users find difficult to parse. This means that not everyone fully understands and recognizes what rights they are giving up by using these technologies, nor do they clearly recognize the rights they are giving these corporations to use their personal data to sell advertisements or reveal trends about their lives and preferences as citizens and consumers.

Such trends become ethically problematic when users do not possess a clear understanding of how their data are being used or when demographic data are used for morally questionable purposes. Oftentimes, aspects of our identity or data describing our lifestyles are embedded within digital hard drives and databases without us even knowing. We have conversations with friends on a social media site and then we notice the creepy intrusion of advertisements that seem oddly in tune with those conversations. For example, there is startling, albeit somewhat anecdotal, evidence that suggests even the microphones installed in our digital technologies are collecting information about our lives that can be used to sell us more things in more specific ways. Consider this BBC article about the types of targeted advertisements that Facebook users experienced (BBC.com, 2017):

- An engaged couple experienced wedding ads after they bought the engagement ring, but before they told anyone.
- A job hunter who recently left his prior employer and jokingly mentioned he might end up in Starbucks next was provided with information about an employment event for that very company.
- An individual had a conversation with a friend about outdoor storage and then was surprised when a storage shed advertisement appeared in her Facebook newsfeed.

What is particularly surprising about these examples is not the fact that targeting advertisements were so efficiently delivered to these devices' owners, but rather that they were delivered without any deliberate postings on social media from the users. In other words, the advertisements were served up in the background based on audio of real-world conversations captured through their devices' microphones! Nichols (2018) investigated this phenomenon after experiencing a similar story in which he mentioned recent trips to Japan to a friend and then was served pop-up advertisements in Facebook about tickets to Tokyo. He interviewed a cybersecurity expert, Dr. Peter Henway, to learn more about smartphones and how they listen to their users. What he learned from Dr. Henway is that there are specific triggers that indicate to the smartphone to pay attention and record the conversation. "Alexa," "Hey Siri," and "Okay Google" are common triggers for modern voice-activated digital devices, for example. However, Dr. Henway also noted the following:

> In the absence of these triggers, any data you provide is only processed within your own phone. This might not seem a cause for alarm, but any third-party applications you have on your phone—like Facebook, for example—still have access to this "non-triggered" data. And whether or not they use this data is really up to them.
>
> *(Nichols, 2018, para. 3)*

This seemed suspicious to Nichols, so he investigated further by repeating phrases that he thought would be likely trigger candidates—phrases like "I'm thinking of going back to uni" and "I need some cheap shirts for work" (Nichols, 2018, para. 7). What he found was an immediate, overnight result. Suddenly, he was seeing ads for courses at various universities as well as certain brands of clothing in his Facebook advertisements. Nichols learned through his conversations with Dr. Henway that although no social media companies were directly selling data to advertising companies, they *were* providing companies with access to demographic groups that might be interested in certain products.

Beyond the intuitive creepiness of intrusive advertising many may feel, there are a number of ways this embedded demographic data might be morally problematic, as evidenced through the Cambridge Analytica scandal where demographic data harvested by a company owned by a major right-wing donor was misused for political gain (Confessore, 2018) or the use of "microtargeting" or "weaponized ad technology" to attempt to influence very specific groups of Internet users (Singer, 2018). In one particularly egregious example, a Russian-linked organization discouraged African-Americans from voting in the 2016 election by spreading memes about disillusionment and the inability of elected officials to represent the interests of nonwhite Americans. Potential voters meeting the algorithm's demographic profile were encouraged not to mobilize and become involved in politics, but rather to boycott the presidential election altogether (Singer, 2018).

Privacy and Anonymity

The examples of weaponized advertisements and micro-targeted demographics lead to an even more fundamental issue in embedded media, which is privacy. Even if we do not choose to carry mobile networked devices around with us everywhere we go, the boundary-dissolving nature of digital media opens up significant access to data we may feel is better kept private. Because digital technologies, sensors, cameras, and databases are ubiquitously embedded in modern society, even those individuals who live their lives mostly disconnected from digital technology are still discoverable in various ways. Consider these examples from Quinn (2015, pp. 227):

> Do you want to know where I live? If you visit the WhitePages.com Web site and type my phone number into the Reverse Phone field, it returns a page giving my name and address. Click on the address and you'll soon see a map of the neighborhood around my house.
>
> Spend a few seconds more, and you can learn a lot about my standard of living. Go to Zillow.com and enter the address that you just learned from WhitePages.com. Zillow dutifully returns the estimated value of my house, based on public records that document its size, its assessed value, and information about recent sales of other homes in my neighborhood. Click on the Street View tab and you'll see a photo of my house taken from Google's camera-equipped car as it passed down my street.

Quinn (2015) continues to explain that you can learn even more about his life if you use social media and happen to be friends with one of his Facebook friends online. You can see pictures of his family and learn more about his hobbies and activities. While there are ways to restrict such portals to only our immediate friends, or not to use them in the first place, the example serves to highlight one of the fundamental shifts in how we think about privacy. Traditionally, philosophers have discussed privacy in terms of access, defined as either "physical proximity to a person or knowledge about that person" (Quinn, 2015, p. 229). Using this definition, we could argue that in general, privacy in society at large has been reduced because there is so much information describing individuals that is embedded into our databases and networks—information pertaining to our homes, our standards of living, and our lifestyles, as evidenced in the example above. And we can further see that things we might choose to make private in the real world—such as our detailed preferences about specific types of media content, our political or religious affiliations, or even our ethnicities—may be shared behind the scenes so that Facebook, Google, and other technology companies can make money selling advertisements. With advances in artificial intelligence and facial recognition technologies, this information might not even need to be provided by us, but rather extracted from pictures we are "tagged" in using sophisticated algorithms and image filters to make guesses about our ethnicities, clothing preferences, or any other number of data points that might be extracted from visual imagery to a reasonable degree of accuracy. There are already software products on the market, such as Microsoft's Face API, that do a reasonably good job at determining one's age and current emotions based solely on a photograph from a live camera (Microsoft, 2018).

Throughout this book we have been stressing the experiential dimension of ethical decision making. In Chapter 1, for example, we discussed digital literacy as a mechanism for understanding the difference between the conceptual and the procedural. This pertains to notions of embeddedness and privacy, too. If we think of the word "embeddedness" to mean something akin to "placed within," then there is both the question of the data themselves (e.g., how is data stored, is data appropriately protected, etc.) as well as how those data are accessed (e.g., the procedures necessary to access and do things with data). In the Zillow case we discussed above, for example, the intent is purportedly not to allow any random website visitor to determine the social standing of his coworkers by valuating their real estate, but rather to create a usable and scalable digital system to aid in the buying and selling of residential properties. However, with the appropriate digital literacy skills, one can turn information sources like Zillow into tools for digging into information that many users would rather be kept private.

What is worrisome about this particular example is that a homeowner does not have the ability to "opt in" to this service; the data is available online to the general public whether they want it to be or not. While one might argue that Zillow is merely pulling data from public records that already exist (and that in fact existed long before Zillow as a company was even formed), the distinction is that Zillow's system is more usable, integrated, accessible, and embedded in the everyday web searching patterns of millions of individuals. That makes the information it provides more useful to potential home buyers and sellers, but also more dangerous for those individuals doing neither of those things who would prefer to keep their home values private.

A feature of embeddedness related to privacy—one that is somewhat counter-intuitive at first—is anonymity. One might think that living in a digital world makes one more personally identifiable than ever before due to the digitization of identity and the availability of one's personal activity information, as discussed earlier in this chapter. This is true to some extent, as we discussed above with respect to privacy, but it is also true that one can often hide behind anonymity in digital environments. Internet Service Providers (ISPs) may provide records of IP addresses and other identifying information to authorities upon request in criminal cases, but to the average person trolling his local newspaper online, this is not something that he will be very concerned about. And even those data requests can be mitigated with knowledge about anonymous VPNs, Tor browsers, firewalls, and other Internet activity–hiding technologies. Or users may simply drive around a neighborhood until they find an unsecured Wi-Fi access point or visit a publicly accessible Wi-Fi access point or computer lab. This allows them to pursue online activities in a fashion where their anonymity is reasonably assured, if not fully guaranteed. In these situations, even if an IP address is requested and delivered to authorities, it will be someone else's equipment that registered that address on the network.

It is not surprising, then, that in discussions of ethics and digital technology, anonymity is often one of the features highlighted as leading to unique ethical problems that we may not consider in other contexts (Vanacker & Heider, 2016). And there are other complications, too. For instance, when we participate in online transactions, some aspects of our identities are also embedded in the digital, taking on the same malleable and programmable characteristics as the medium itself. It is possible that our virtual identities, or some dimensions of them, may be quite different from our real identities. To some, this may be a powerful and freeing feeling, and Sherry Turkle writes at length about individuals who felt this way in *The Second Self*. However, the anonymity afforded by embedded identity can also be morally problematic, as evidenced by Internet trolling, hate speech, and offensive discourse in online communities.

Finally, we may also think of embeddedness and identity in terms of how anonymous content is embedded into non-anonymous content, creating a troublesome power differential in online discourse. For example, gender and racial issues are significant in online content, particularly in digital journalism. A recent study commissioned by *The Guardian* found that after analyzing 70 million comments left on its website since 2006, eight of the ten most abused writers were women and the other two were black men (Gardiner et al., 2016). In this sense, abuse is embedded into the context of the original articles, since the comments appeared below the articles and, by extension and by virtue of their proximity and organization, became part of the context of the original discussions. While the original article had an author's real name in the byline, the anonymous posters spewing vitriol were protected by anonymity and hid behind fake names or deliberately misleading online identities.

Since we have stressed the importance of action throughout the book, we might ask ourselves: What can we do about the problematic issue of digital anonymity? As you know by now, this is a complicated problem. Because of the scale and reach of the Web, it is often impossible, or at least economically unsustainable, for human beings to police online communities and ensure civil discussions are taking place in our online spaces. The government certainly cannot keep up with this. The Department of Justice, for example, has one single person to keep track of the estimated 784 non-Islamic hate groups in the U.S., according to the Southern Poverty Law Center (Carlson, 2016). This means that unless companies and governments are willing to invest the appropriate resources and human capital necessary to monitor digital environments for offensive or harmful rhetoric, we will see several possible outcomes. One outcome is that the content will simply not be monitored and trolls and bigots will reign free in digital space. Another outcome is that the monitoring will be turned over to computer algorithms, which, as we have discussed throughout the book and particularly in Chapter 5, can be problematic for many reasons. And a third option is that the identification of offensive rhetoric can be crowdsourced to the users themselves, allowing users to "flag" offensive content. If enough users flag a specific instance of content, that may then be forwarded to an algorithm or a human being for subsequent analysis. In all these scenarios, there are ethical challenges to be overcome. Humans are subjective and we have unconscious biases and we make mistakes. Algorithms are still terrible at certain types of tasks, like detecting sarcasm and nuance or applying common sense. And communities of users are not necessarily "right" in their judgement just because a large group of them believe the same thing. Hate groups online are testament to that.

Despite the challenges involved with policing digital content, countries are beginning to develop legislation that mandates it must be done. For example, a recent law passed in Germany requires social media companies to remove illegal, racist, or slanderous content within 24 hours or risk facing hefty fines (some as large as $57 million) and a Canadian court ruling "ordered content that violated Canadian law should be deleted globally rather than just for Canadian users" (Wadhwa & Ng, 2017, para. 2). As the authors of a *Wired* article point out, however, this legislation essentially moves the responsibility of content enforcement to these large technology companies such as Twitter and Facebook (Wadhwa & Ng, 2017). This is troublesome because, as these authors note, one's freedom of expression may vary greatly depending on geographic location, but the technology enabling these online conversations to take place transcends any geographic borders. Further, the companies themselves are not structured to serve as legal or political bodies. Because of this, the algorithms and policies they implement to police their content "may use rules to police content that lack the clarity, protections, and appellate procedures that the rule of law requires" (Wadhwa & Ng., 2017, para. 7). Enforcing moderation globally without the local perspective and legal protections afforded to physical spaces in the community is both technically challenging and morally problematic. However, this is simply another side effect of the geography-dissolving nature of the Internet.

Organizational Values

Another relevant aspect of embeddedness is its role in corporate or organizational values as evidenced through design and development decisions. As Couldry, Madianou, and Pinchevski (2013) point out in *Ethics of Media*, media technologies are critically important—not only because of their "sheer pervasiveness" and their status as "centralized institutions" (2013, p. 1) in and of themselves, but also and perhaps more importantly because they mediate our relationships with one another. Unless we are designing our own hardware and writing our own programming code, we are going to implicitly adopt many of the same perspectives about technology and how we use technology to interact with one another as the organizations and companies that design the digital products we use. In this way, organizational values are embedded into their products and then we gradually begin to see the world through their eyes, whether we realize it or not. This is troubling when large technology organizations are moving so quickly that they overlook or minimize potential ethical problems in order to reach their markets in the most efficient manner possible and retain their competitive advantage. Many of these companies also lack chief ethics officers to help them consider thorny problems with their technologies that transcend the legal matters their attorneys deal with (Swisher, 2018).

In an opinion piece titled "Who Will Teach Silicon Valley to be Ethical," journalist Kara Swisher bemoans the number of ethical dilemmas currently facing high tech organizations, problems that do not seem to be slowing these companies down from their forward momentum (2018). For example, Swisher outlines several troubling scenarios, many of which prominently feature embeddedness in one way or another. Swisher's commentary describes events unfolding in the fall of 2018, including these examples:

- The situation in which large amounts of Saudi funding (billions of dollars) are directed toward U.S. technology companies while at the same time Saudi Arabian leaders are publicly implicated in the beheading of a journalist.
- The launch of a new embedded video device introduced by Facebook, called Portal, that joins other voice assisted devices currently on the market for hands-free calling. The technology can collect data about who you call and the apps you interact with in order to serve advertisements on Facebook.
- The fact that the technology company Google waited six months to inform users of its mostly defunct social network Google Plus that their data had been compromised.

Each of these examples can be considered through the lens of embeddedness. For example, the Saudi Arabian funding examples show there is a willingness of Silicon Valley companies to overlook broader political or moral dilemmas within the Saudi Arabian government if the money continues to flow. Here, embeddedness emerges as a double standard, in that different standards of morality are embedded in the political operations of a country than are embedded in its economic operations. To be fair, this situation of money foregrounding morality is true of many other countries outside of Saudi Arabia, including the U.S.

The Portal example reveals troubling patterns in consumer privacy like those we discussed earlier in the chapter. Previously, we discussed examples in which advertisements can be generated based on audio conversations recorded by the microphones in their users' mobile devices. Portal continues the trend where users become embedded in the products they use, allowing advertisements to be more stealthily served to potential consumers as they go about their daily lives and talk with friends and family online. Here again, the convenience of such devices is what persuades users to give up significant amounts of their privacy to use them.

Finally, the Google example shows that the eventual loss of privacy and security have now become embedded in our expectations about technology companies; it has become the norm rather than the exception. Google is by no means the only company that has had its user data stolen by hackers; the company joins a long list of other technology giants including Yahoo, Adobe, Sony, eBay, and JP Morgan Chase (Armerding, 2018). We discussed the data breach at Equifax at length in Chapter 4. Joining these large companies are the thousands of other smaller organizations that also experienced data breaches and may or may not have reported them. The fact that this happened with Google is not what is surprising here. What is surprising is the seemingly cavalier attitude about waiting so long to report it, especially for a company known at one time for having "Don't Be Evil" as a motto within its corporate code of conduct. This delayed disclosure is, as Swisher points out, highly concerning, but it shows how commonplace these expectations about (poor) security and (the lack of) privacy have become in our ideas about digital technology.

Another way of considering organizational values is to think about the organization's "brand track record," or the core values as reflected in decisions and activities made by the organization over time. As Urde (2009) describes it, such a track record is made up of:

> Promises rooted in the organization, which are also perceived and appreciated by customers and non-customer stakeholders over time. It is an emerging pattern of proven values and promises forming a contract between the organization and the outside world. A track record reflects continuity, influences the customer's expectations, and is part of a corporate brand's identity (p. 620).

Consider Volvo, an automobile company that initially held three core values as central to their corporate philosophy: quality, safety, and environment (Urde, 2009). One can trace specific decisions made by Volvo over time to verify whether the company "put their money where their mouth is" in terms of their brand track record. Specific design decisions, such as the decision to invest resources in areas such as padded dashboards, additional seat belts, side air bags, and rollover protection all reinforced a strategy to hold safety concerns paramount in their design and manufacturing of vehicles. Similarly, their commitment to the environment is reflected in other design decisions, such as "hybrid engines for heavy-duty vehicles" (Urde, 2009, p. 625) and their commitment to quality is evident through products that generally scored high in reliability measures over time.

The cases of Volvo and other modern automobile companies are especially interesting from an ethical perspective because as modern automobiles become more complicated, digital electronics and embedded computing devices become more central to their operation. In fact, some have described modern automobile manufacturing as something akin to software development and there are essays arguing that software is going to soon be even more central to the automotive industry, moving "towards a 'technology stack' that in many ways resembles what we have seen occur in the PC industry" (Clarence-Smith, n.d., para. 12). When this embedded digital technology resides in portions of the automobile that had previously been governed by analog controls, such as emissions and exhaust, it becomes possible to digitally manipulate the resulting controls or the data from those controls in various ways.

The diesel emissions scandal that cost Volvo's competitor, Volkswagen, millions of dollars and significantly damaged their reputation is one example of this. In a series of events now known as "Dieselgate" (Kerler, 2018), it is alleged that the company modified their vehicles to allow them to pass U.S. emissions tests and then attempted to keep that information hidden from investigators (Woodyard, 2018). The automobile company Audi has also been implicated in "developing and installing illegal software in 11 million diesel cars in order to trick emissions tests" (Kerler, 2018, para. 2) which ultimately prompted investigations from governmental agencies in the U.S. and around the world (Kerler, 2018). Perhaps Volkswagen would have taken this ethical misstep even without the integration of digital technology into their vehicles. However, this example is instructive not only because digital technology made it easier for the company to manipulate emissions readouts, but also because digital information flows made the impact of that misstep global and swift.

As these scenarios illustrate, the brand track records of modern automotive companies can, in our new digital landscape, be evaluated not only subjectively, but also quantitatively. Decisions and claims can now be interrogated based on data from these embedded digital devices and the embedded sensors connected to them. In terms of brand track record, this allows both consumers and government regulators to trust, but verify, claims of efficiency and operation. This is recognized in the evolving philosophies of some of these modern automobile manufacturers, where short term ethical stumbles may pose long term consequences for their companies. For example, Volvo's new core values have been updated and are now focused on quality, environment, and *ethics* (Volvo, 2018). For "ethics" to replace "safety" in that original value statement speaks volumes, particularly when Volvo is probably most known for its commitment to safety. Then again, this existing reputation is likely what allows them to shift their focus in a new direction, since their safety image is already well established.

Sometimes, organizational values impact digital culture in a seemingly positive way but then problems emerge later. For example, Apple Inc.'s quest for simplicity in its products is well documented in usability and design circles (e.g., Maeda, 2006). This reduction of complexity, particularly at the user interface level, has allowed entirely new audiences of users, such as very young children and the elderly, to experience sophisticated computing devices that were formerly too complex for many of them to use. Instead of complicated buttons and software filled with layers of menus and options, Apple instead chose to develop hardware devices and operating systems for its mobile devices that could be operated using natural human motions: swipes, pinches, and presses. However, even this seemingly virtuous philosophy of design simplicity is not so black and white.

It seems like an altruistic decision on the part of Apple leadership and developers, and perhaps that plays a part in their business and design decisions. However, devices easy enough for toddlers to use also now become fertile territories for dubious marketing tactics (and opportunities to indoctrinate lifelong Apple users, too). If not configured with proper security, they may also expose young children to mature subject material on the Internet. Similarly, elderly populations using the devices will now experience new types of scams and marketing strategies designed to lure less sophisticated digital technology users away from their material assets. There are many reasons why this is case, but one major catalyst is the fact that the elderly population of computer users is rapidly growing and provides an enticing group of targets for unethical scammers and businesses of low moral integrity.

At some point organizations will decide where exactly they fall on the "usability versus security" continuum and how much risk they are willing to take on for their users. Much has been written about the tradeoff between usability and security in computing devices and the topic is frequently discussed in various computing contexts (e.g., Jackson, 2017). Seemingly, very intuitive and usable devices are not as secure as their less open and accessible peers. As a simple example, a website that allows its users to select three digit passwords is very usable, since it requires minimal typing and three digit passwords are easy to remember. But it is not very secure, since passwords that short are easily hacked. Conversely, more secure devices are generally inconvenient to use. As an example, our home institution recently implemented a multifactor authentication system to enable logging into the university's computer databases. A system of this sort requires more than one method of authentication to grant access to secure resources such as protected webpages and electronic documents. So, a user might need to enter a password into a website, but also provide a security code that the system sends as a text message to the user's mobile device. Entry is then granted if both the password and the security code are valid. At universities and other large organizations that rely on electronic databases for their daily operations, employees do this frequently to access human resources records and other types of restricted data. While the system is certainly more secure, it is also more of a hassle to log into since it now requires an additional code to be entered. Because the code is texted to a mobile phone or another device not associated with the computer we are logging in from, if we do not have our phones handy, we cannot log in. Thus, security comes at a very real expense to usability.

Apple chose to address some of the potential security concerns in their programs by making their software highly controlled. The process to have an "app" approved on the Apple iOS App Store is rigorous. After creating apps, developers must submit their software to Apple for review before the product can be downloaded from the App Store and installed on users' devices. The company then checks to determine "whether they are reliable, perform as expected, and are free of offensive material" (Apple, 2018). Many of Apple's guidelines appear virtuous and protective of their user base. For instance, many apps are rejected because the apps are incomplete, buggy, or of dubious value (e.g., classified as spam). However, Apple may also reject applications for reasons such as "substandard user interfaces" and "not enough lasting value." These two categories are subjective enough to allow the company to impose their own organizational interpretations of these words to reject software that they do not see as fitting into the embedded ethos of Apple Inc.

Cultural Change and Relationships

There are also ethical implications to the more physical practicalities of embedded digital technologies. In this case, we mean the nature and extent to which digital computing devices are embedded in our personal and professional lives. A report by the McKinsey Global Institute (2017) predicts massive shifts in our global workforce by the year 2030. Due to the rise of robotics and automation, McKinsey estimates that between 400 and 800 million of today's jobs will be automated by 2030. This belief is congruent with the expectations of everyday people, with a recent Pew Research Center poll indicating that large majorities of people in countries around the world believe that robots will be doing much of the work of humans in 50 years (Washington Post, 2018). In this sense, then, it is not only digital technology that will be embedded within our lives, but also, that the reverse is true. In other words, it is also the case that at least in the workplace, there is a very good chance that we will need to embed *ourselves* into environments increasingly populated by automated digital technologies such as hardware-based robots, software-based "bots," expert systems, and machine learning systems.

In our personal lives, this embeddedness is highlighted through the emergence of increasingly connected networked devices and the Internet of Things (IOT). Many of us may hear this term and associate IOT with gimmicky household appliances like networked smart toasters, refrigerators with built-in databases and cameras to tell us when we are out of eggs, and online coffee makers and microwaves with support for voice commands (as though you don't need to walk over to these devices anyway, to grab your food or coffee). However, even if one is not an early adopter of such gadgets, he or she may still be affected by the IOT, especially if he or she uses smartphones and mobile devices. Consider how networked digital technologies are gradually replacing their analog counterparts. Greengard (2015) argues this is already happening, writing, "Conventional cameras and film have largely disappeared, standalone video and audio devices are disappearing, paper maps are vanishing, landline phones are on the way to going extinct, and traditional books and magazines are becoming a chapter in history" (loc. 224). We instead rely upon digital devices that can serve these functions; in fact, many devices like modern mobile phones can now do *all of them*. Anyone who owns an Apple or Android smartphone can record video and audio, find the closest route to a friend's home for a party, read books on specialized eReader software, update and read from their social media feeds, and text with friends individually or in groups. And, if all else fails, they can still make phone calls, too.

We are also online and connected to networked devices more than ever before. As of late 2015, the Pew Research Center reported that American adults were going online every day with 20 percent online "almost constantly" (Wachter-Boettcher, 2017). This means our lives are networked so thoroughly that digital technologies are affecting not only our individual selves, but also our relationships with other people in the physical world. For example, the young adults interviewed by Ito and colleagues (2010) spoke of the value of social media both in forming and maintaining social relationships and friendships with peers. Because of the networked nature of the Internet and the massive speed at which information can be exchanged through mobile devices and social media, our relationships too are mediated and shaped by the technologies we use. And because of the power of scale and the ability of computers and databases to easily extract similar groups of data, we also have unprecedented opportunities for mobilizing very specific groups of people for various purposes.

When aspects of our lives are codified, digitized, and stored in databases, computers can use this information to categorize us into both similar and dissimilar groups. This data can then be used for different reasons, good and bad. At worst, we get echo chambers, where opinions seem immutable and where constant feedback loops reinforce already strong ideas about politics, religion, or culture. Or we package our communities into demographic clusters that can be sold to advertising companies or manipulated by disinformation campaigns on social media, as we discussed earlier in this chapter.

However, such embedded communities can also be quite positive, or even therapeutic. Consider the case of Roxane Gay, a writer who experienced a terrible sexual assault as a child and was unable to talk about this assault with her parents, friends, or community. She was able to use the Internet not only for companionship, but also to work through her feelings and write about what had happened. An audience in the real world was difficult to identify, but online, it was sadly too easy to find communities of individuals with similar experiences. In her memoir *Hunger,* Gay writes,

> When I got home at night, I generally went straight to my computer, where I wrote story after story, mostly about women and their hurt because it was the only way I could think of to bleed out all the hurt I was feeling. I frequented newsgroups and chat rooms for survivors of sexual assault. Though I couldn't tell anyone in my real life what had happened, I unburdened myself to strangers on the Internet. I blogged, mostly about the minutiae of my life, hoping, I think, to be seen and heard. I loved and craved the freedom of being online and being free from my life and my body.
>
> *(Gay, 2017, loc 929–959)*

As Gay's experience shows, one positive side effect of the mass embeddedness and ubiquity of networked media is the possibility of connecting immediately to communities you could not easily assemble in your local neighborhood or even your local city. Because of the scope of the Internet and its widespread adoption throughout the world, you can extract a subset of almost any group you can think of from the estimated 3.6 billion users of the Internet (Statista, 2017). Even health conditions that are for the most part unwelcome or taboo subjects for discussion between friends and even among family members find a home on the Internet with individuals forming communities of interest around conditions like irritable bowel disease, Crohn's and colitis, and the (perhaps surprisingly large) community of people living with intestinal diversions such as colostomies and ileostomies. Where discussions about bodily waste are even discouraged or frowned upon in physicians' offices, people who are members of discussion groups online feel free to discuss the problems, successes, surgeries, pain, and ideas for effectively living with such surgical changes to their bodies that were once not discussed in "polite" company at all. On the other hand, there are trolls and bullies in communities like these, and many of the discussion boards for health-related conditions employ or have trusted board members as volunteer moderators to keep the trolls and bullies at bay so that friendly and productive discussions can continue. Many people who are members of such online groups have found a community of like-minded (and like-bodied) others to whom to express their emotions and their concerns without undue fear of negative social reactions or the unfortunate ostracism that comes about as a result of lack of understanding.

We can see yet another example of this type in the private servers found in videogames. On these servers, players can share supportive experiences and perspectives with other similarly minded gamers. Yet, they can also face bullying, hate-speech, bigotry, aggression, manipulation, and shaming from within those same communities across those same servers. The concept of embeddedness need not be restricted to the embeddedness of digital technology in our lives. It is also the case that digital technology makes possible the creation of relationships and communities that could not have existed without digital technologies and are more dynamic than one may imagine.

Even our relationships with ourselves can change due to digital technologies, particularly in anonymous contexts. Virtual identities, as we discussed earlier in this chapter, are much more malleable and transformable than the identities we project in our face-to-face interactions. When these identities are also anonymous or fictional, it is easy to take risks, too, since the worst that can happen is one virtual identity needs to be scrubbed out and replaced with another. Digital identities can be liquidated as easily as stocks and bonds. Some individuals may try out new personae on a whim, perhaps by posting aggressive and uncharacteristic comments on online news articles or by enacting a new gender in a massively multiplayer online role-playing game. We may reflect upon what we consider to be poor moral choices, leading to feelings of sadness or regret, or we may find new opportunities for empathy that previously were outside of our awareness. In these cases we may think of ourselves differently due to the way we acted when we "were" digital. Such changes in ethos are mediated by self-reflection and occur when our identities and behaviors in online communities are different from our "analog" existence.

Special Considerations

Finally, it is worth considering that embedded technologies may also have implications for specialized audiences and stakeholders, such as users or audiences from specific demographics or age groups. Children and older adults are two good examples of groups of users that may require special consideration. While embeddedness is not the only property of digital media that influences these audiences, it does raise some interesting issues and examples.

For instance, there are many positive benefits of digital technology that are taken advantage of by elderly users. The term "elderly" is admittedly difficult to define in terms of a specific age range (Weeks, 2013), so here we use it loosely while recognizing that it may mean different age ranges to different people. In any case, our digital devices bring communication, navigation, entertainment, and business improvements to older people in the same ways they do for younger populations. However, elderly populations are also particularly vulnerable to negative effects, such as being scammed online. Twenty percent of elderly consumers report being the victim of financial abuse, much of which now occurs electronically, with an overall cost to elderly consumers of $36.5 billion (Olson, 2018). Coombs (2014) explains that older women are prime targets for electronic fraud because women tend to live longer than men, leading to situations in which they may be managing their finances for the first time and may be more likely to show compassion due to prior experience as mothers or caregivers. However, she also notes that men are highly susceptible to electronic abuse, falling victims to scams such as the "sweetheart scam" where younger women befriend and financially exploit older men after the loss of their spouses. These elderly men and women are susceptible to electronic fraud because of their respective wealth as compared to younger demographics (Coombs, 2014), making them juicy targets for online scammers. Many also have good credit due to the lengths of their credit histories and are reluctant to report crimes due to fear of harassment (Nino, Enstrom, & Davidson, 2017).

Situations like this offer new scenarios for considering embeddedness. Older adults are embedded into particular types of familial relationships and friendships and as they age they begin to lose close family members, partners, spouses, and lifelong friends. Technology can play a role in providing these groups with newly embedded virtual communities, but they can also allow the exploitation of behaviors and activities that have not yet adjusted to the digital age. The ability of individuals to embed their activities with anonymity, as discussed earlier in this chapter, creates a situation in which trust can be manipulated. It is relatively easy for shady individuals and companies to take advantage of the limited technical knowledge of elderly end-users who are often still becoming accustomed to digital technologies.

A 2018 article in the *New York Times* explains that there has been a sustained increase in financial fraud against the elderly in Maine (Olson, 2018). After one case in which retired legal secretary and widow Dawn Shaw was robbed using fraudulent automatic electronic withdrawals, her daughter encouraged Maine state officials to begin a pilot program that would enable bank employees to spot suspicious activities. Employees and banks would then be provided with greater protection from legal liability (e.g., being sued by clients or penalized by regulators) for reporting their suspicions. The resulting Senior Safe Act program, which became law in Maine in May of 2018, was a resounding success. It led to a significant reduction in these types of crimes in Maine and to a state senator eventually introducing legislation to take this program to a national level.

Next Up

These last few chapters have explored three properties of the digital—distributedness, procedurality, and embeddedness—that we think have important ethical implications. The next chapter, introducing Part 3 of the book, focuses on problems of empathy and desensitization as two of these implications. Flows of digital information have the potential to enhance the ways we connect or empathize with others. But those same flows can also desensitize people to events and states of affairs that would normally trigger an empathetic response. We will review examples that consider the impact of the digital on the ways we connect with others.

References

Apple. (2018). App review. Retrieved from https://developer.apple.com/app-store/re view/.

Armerding, T. (2018, January 26). The 17 biggest data breaches of the 21st century. *CSO.* Retrieved from https://www.csoonline.com/article/2130877/data-breach/t he-biggest-data-breaches-of-the-21st-century.html.

BBC.com (2017, October 30). Is your phone listening in? Your stories. *BBC.* Retrieved from https://www.bbc.com/news/technology-41802282.

Carlson, C.R. (2016). Hashtags and hate speech: The legal and ethical responsibilities of social media companies to manage online content. In Vanacker, B. & Heider, D. (Eds.), *Ethics for a digital age* (pp. 123–140). New York, NY: Peter Lang.

Clarence-Smith, T. (n.d.). How software will dominate the automotive industry. *Toptal Insights.* Retrieved from https://www.toptal.com/insights/innovation/how-software -will-dominate-the-automotive-industry.

Confessore, N. (2018, April 4). Cambridge analytica and Facebook: The scandal and the fallout so far. *The New York Times.* Retrieved from https://www.nytimes.com/2018/0 4/04/us/politics/cambridge-analytica-scandal-fallout.html.

Coombs, J. (2014). Scamming the elderly: An increased susceptibility to financial exploitation within and outside of the family. *Albany Government Law Review, 7,* 243.

Couldry, N., Madianou, M., & Pinchevski, A. (Eds.). (2013). *Ethics of media.* Basingstoke, UK: Palgrave Macmillan.

Ferguson, C. J. (2015). Do angry birds make for angry children? A meta-analysis of video game influences on children's and adolescents' aggression, mental health, prosocial behavior, and academic performance. *Perspectives on Psychological Science, 10*(5), 646–666.

Floridi, L. (2014). *The fourth revolution: How the infosphere is reshaping human reality.* Oxford, UK: Oxford University Press.

Gardiner, B., Mansfield, M., Anderson, A., Holder, J., Louter, D., & Ulmanu, M. (2016, April 12). The dark side of Guardian comments. *The Guardian.* Retrieved from https://www.theguardian.com/technology/2016/apr/12/the-dark-side-of-guardian-comments.

Gee, J. P. (2007). *What videogames have to teach us about learning and literacy.* Revised and updated edition. New York, NY: Palgrave Macmillan.

Greengard, S. (2015). *The internet of things.* Cambridge, MA: MIT Press. [Kindle edition].

Ito, M., Antin, J., Finn, M., Law, A., Manion, A., Mitnick, S., ... Horst, H. A. (2010). *Hanging out, messing around, and geeking out: Kids living and learning with new media.* Cambridge, MA: MIT Press.

Jackson, B. (2017, October 19). Security versus usability: Overcoming the security dilemma in financial services. *Microsoft industry blogs.* Retrieved from https://cloudblogs.microsoft.com/industry-blog/industry/financial-services/security-versus-usability-overcoming-the-security-dilemma-in-financial-services/.

Kerler, W. (2018). You thought Dieselgate was over? It's not. *The Verge.* Retrieved from https://www.theverge.com/2018/9/18/17876012/dieselgate-volkswagen-vw-diesel-emissions-test-epa-german-auto-industry-mercedes-benz-bmw.

Lessig, L. (2004). *Free culture: How big media uses technology and the law to lock down culture and control creativity.* New York, NY: Penguin.

Maeda, J. (2006). *The laws of simplicity.* Cambridge, MA: MIT Press.

McKinsey Global Institute (2017, December). Jobs lost, jobs gained: Workforce transitions in a time of automation. Retrieved from https://www.mckinsey.com.

Microsoft. (2018). Microsoft Face API. *Microsoft Azure.* Retrieved from https://azure.microsoft.com/en-us/services/cognitive-services/face/.

Nichols, S. (2018, June 4). Your phone is listening and it's not paranoia. *Vice.com.* Retrieved from https://www.vice.com/en_uk/article/wjbzzy/your-phone-is-listening-and-its-not-paranoia?utm_source=vicefbuk.

Nino, J.-R., Enström, G., & Davidson, A. R. (2017). Factors in fraudulent emails that deceive elderly people. In J. Zhou & G. Salvendy (Eds.), *Human aspects of IT for the aged population: Aging, design, and user experience* (pp. 360–368). Cham, Switzerland: Springer.

Olson, E. (2018, August 18). Helping banks flag fraud against seniors. *The New York Times.* Retrieved from https://www.nytimes.com/2018/08/18/business/banks-financial-fraud-seniors.html.

Quinn, M.J. (2015). *Ethics for the information age.* (6th ed.). Boston, MA: Pearson.

Singer, N. (2018, August 16). 'Weaponized ad technology': Facebook's moneymaker gets a critical eye. *The New York Times.* Retrieved from https://www.nytimes.com/2018/08/16/technology/facebook-microtargeting-advertising.html.

Smith, S., Ferguson, C., & Beaver, K. (2018). A longitudinal analysis of shooter games and their relationship with conduct disorder and self-reported delinquency. *International Journal of Law and Psychiatry, 58,* 48–53.

Snider, M. (2018, October 1). Study confirms link between violent video games and physical aggression. *USA Today.* Retrieved from https://www.usatoday.com/story

/tech/news/2018/10/01/violent-video-games-tie-physical-aggression-confirmed-study/1486188002/.

Statista (2017, July). Number of internet users worldwide from 2005 to 2017 (in millions). *The Statistics Portal*. Retrieved from https://www.statista.com/statistics/273018/number-of-internet-users-worldwide/.

Swisher, K. (2018, October 1). Who will teach Silicon Valley to be ethical? *The New York Times*. Opinion. Retrieved from https://www.nytimes.com/2018/10/21/opinion/who-will-teach-silicon-valley-to-be-ethical.html.

Turkle, S. (2005). *The second self: Computers and the human spirit*. 20th anniversary ed. Cambridge, MA: MIT Press.

Urde, M. (2009). Uncovering the corporate brand's core values. *Management Decision*, 47(4), 616–638.

Vanacker, B. & Heider, D. (Eds.). (2016). *Ethics for a digital age*. New York: Peter Lang.

Volvo. (2018). About Volvo cars. Retrieved from https://group.volvocars.com/sustainability.

Washington Post. (2018, September 13). People around the world think that robots will soon take most human jobs - and that people will suffer. Retrieved from https://www.washingtonpost.com/world/2018/09/13/people-around-world-think-that-robots-will-soon-take-most-human-jobs-that-people-will-suffer/.

Wachter-Boettcher, S. (2017). *Technically wrong: Sexist apps, biased algorithms, and other threats of toxic tech*. New York, NY: WW Norton & Company.

Wadhwa, T. & Ng, G. (2017, July 31). Tech companies policing the web will do more harm than good. *Wired*. Retrieved from https://www.wired.com/story/tech-companies-policing-the-web-will-do-more-harm-than-good/.

Weeks, L. (2013, March 14). An age-old problem: Who is 'elderly'? *NPR*. Retrieved from https://www.npr.org/2013/03/12/174124992/an-age-old-problem-who-is-elderly.

Woodyard, C. (2018, May 3). Ex-Volkswagen CEO indicted in alleged plot to rig diesel cars to pass U.S. emissions tests. *USA Today*. Retrieved from https://www.usatoday.com/story/money/cars/2018/05/03/volkswagens-former-ceo-five-others-indicted-diesel-car-scandal/578755002/.

PART 3

Implications of Digital Ethics

7

DIGITAL RELATIONS AND EMPATHY MACHINES

Empathy is essential for ethics literacy. Becoming ethically literate requires understanding others' perspectives as a prerequisite to the work of cultivating ethics sensitivity, reasoning through the logical implications of concepts and theories, or finding the motivation to act on those reasons. Philosophical ethicists have, historically, privileged that rational side of ethics: focusing on arguments and conceptual development and leaving the "squishier" human interaction stuff to others like, for instance, qualitatively oriented psychologists. In the last couple of decades, philosophers and psychologists have come to intersect in the area of moral psychology. In those conversations we find space for understanding the social, neurological, biological, and affective processes that bump up against even the best efforts to reason, ethically. Those intersecting skills define empathy, the capacity for understanding the perspective of others. In the context of digital ethics, empathy is the skill to understanding the *value perspective* of other stakeholders around a particular digital ethical issue. So, in this chapter, we focus on questions of how we relate to and understand other individuals (all, and not only human) in the digital as an essential baseline for thinking about the implications of decisions we make. Cyberbullying is one powerful example of the problem of empathy in the digital.

CASE STUDY: CYBERBULLYING

In early 2018, Panama City, FL middle-school student Gabriella Green hanged herself with a dog leash in her closet. Gabriella had confided in another middle school student, via social media, that she had previously attempted suicide by hanging. "(He) responded by saying something to the effect of, 'If you're going to do it, just do it,' and ended the call," police wrote. "He immediately regretted that statement, and began calling and text-messaging her, but did not receive a response" (Associated Press, 2018).

Shortly after her death, two 12-year-old middle school students were arrested for cyberbullying in connection with her suicide. Both students confessed, acknowledging that their conduct was directed at Green and would result in emotional distress.

School officials called the situation "absolutely tragic" while Gabriella's mother blames those same officials for enabling the situation. Police reported discovering several middle-school children had unrestricted and unmonitored access to social media apps. Police and school officials, in direct reaction to the tragic death, began training and anti-bullying programs at the school for students and parents. Later that year, the Florida legislature approved the "Jeffrey Johnston Stand Up for All Students Act" 1006.147, which expressly prohibits bullying or harassment of any student or employee of a public K-12 educational institution, including cyberbullying (Florida Legislature, 2018).

Despite the digital context of Gabriella's bullying experience, research continues to demonstrate a complex and developing relationship between real-world bullying and online or cyberbullying. For example, a 2012 survey of Canadian youth reported that 25–30 percent of respondents reported experiencing cyberbullying compared to 12 percent who say they had experienced 'schoolyard' bullying (UBC News, 2012). Compare this to a 2017 survey by anti-bullying organization Ditch the Label that reported only 17 percent of respondents experienced cyberbullying compared to 50 percent who had experienced bullying of the 'schoolyard' variety, based on their own definition (Ditch the Label, 2017, p. 21). Conflicting results like these open questions about how cyberbullying has changed, grown, or morphed over time, and the ways in which individuals perceive and understand the relationship between physical and cyber aggression. Yet, we have plenty of evidence that supports the claim that young people are more than twice as likely to self-harm or enact suicidal behavior if they have been victims of cyberbullying (John et al., 2018). No matter how complicated the situation, cyberbullying was certainly bad for Gabriella. What is the moral relationship between cyberbullying and so-called "schoolyard" bullying? How did the digital context of the case effect the conditions of empathy or the outcome? To what extent should law and policy at state and federal levels legislate this digital ethical issue? Does the digital only limit our ability to relate, or can it expand it, too?

The vast and varied literatures on the human ability to understand other person's perspectives share a common view that this idea of "perspective-taking" is a necessary condition of ethical literacy. But despite this general agreement on its importance, there is still a remarkable diversity of opinions on what this capacity is called, how far it extends, and just why it matters. We will first define some key terms and then offer a series of examples about why empathy is ethically relevant to understanding digital ethics. Then we will focus specifically on perspective-taking in the context of digital ethics and juxtapose this key social/ ethical skill against the idea of desensitization (as the antithesis of ethics sensitivity). The core idea or problem in digital ethics we are pointing out in this chapter is that the digital can both serve to connect people more closely together in communities but also threatens to isolate, segregate, and desensitize people to the needs and interests of others. Finally, we will push beyond human-to-human social interactions to discuss the place of empathic desensitization in the digital within discussions of nonhuman interests, too.

The conversations about empathy (or sympathy or perspective-taking or whatever term is used for this capacity to understand another's value perspective) is increasingly relevant for digital ethics, central as it is to digital ethical literacy. For example, virtual reality attempts to build empathy, but many have argued (e.g., Bollmer, 2017) that it fails to foster empathy in a meaningful way by creating virtual/artificial conditions for the transmission of a particular perspective to the participant. Yet, the "technologies designed to foster empathy *presume* to acknowledge the experience of another but inherently cannot" (2017, p. 64). Others have noted, rather obviously, that incorporating these sorts of socio-emotional dimensions in digital education is "hard" because they require not only cognitive but also metacognitive and affective aspects" (Garcia-Perez, Santos-Delgado, & Buzon-Garcia, 2016, p. 3). In a 2016 study Garcia-Perez et al. tried to illustrate this difficulty by applying an existing qualitative assessment scale for studying empathy to open virtual learning environments. They called the result "digital empathy" and found that trainee teachers have a moderate level of it, while only 10.1 percent had what they defined as "optimal" levels (2016, p. 9). So not only is empathy difficult to understand and tricky to define, but it is also particularly hard to transfer in digital environments. Other studies have sought to work out just what the digital does for or to the possibilities of connecting to others. Gordon and Schirra use "empathy" to describe "emotional engagement with character and or space" (Gordon & Schirra, 2011, p. 179), a trait that does not, on their assessment, transfer well from game environments to real environments. They found that research participants were unwilling to talk about the possibility of their digital character influencing their "real" decisions.

We will get a better sense of the implications of empathy for digital ethics through some examples. But first, let us back up and be more careful about how we are using our terms.

We know understanding others' value perspectives is important for ethics. Without it, our ethics could only ever be *egoistic*, or focused on our own rights and consequences. The next important question to consider is just *how* important it is. Philosophers often lean on logical conditions of sufficiency and necessity to determine one concept's relation to another. There are two ways relevant here that this relationship can play out. First, one could argue that empathy is sufficient for ethical decision making. If this were the case, then being empathetic would be all it takes to be make ethical decisions. It is pretty easy to come up with a counter-example to a claim like this—and we only need one to disprove the conclusion of that argument. Imagine that you are the one from the previous example seeing a friend get hit with a Frisbee. And imagine that you feel, even very strongly and accurately, the displeasure or pain that your friend feels. You might then rightly decide getting hit with a Frisbee is a bad thing, and that doing the hitting is an ethically bad act. There are two problems with a view like this.

First, it works only for clear and simple cases of suffering. More complicated cases of ethics, such as those that involve competing values or long-term con-sequences, cannot be resolved by simply experiencing the immediate emotions of another. Imagine that, rather than the Frisbee to the face, your friend has to decide whether or not to blow the whistle on an employer that is releasing beta-software you know to have large security flaws for a widely used piece of hardware. Ethics is far more complicated than just simply immediate pains and pleasures. Second, it takes for granted the difference between *feeling* and *decid-ing*. To merely take the perspective of another person does not necessarily imply acting on that psycho-emotional understanding. Apart from the important dif-ference in perspective, the rational capacity to decide and then the motivational capacity to act also seem to be important parts of ethical decision making. Thus, the other relevant way the logical relationship can play out is that empathy is necessary but not sufficient for ethical decision making. That is, you can have empathy without ethical decision making, but you cannot have ethical decision-making without empathy (and reasoning and motivation, too).

Indeed, that link between empathy, reasoning, and motivation is an impor-tant one. The link between empathy and digital ethics is the link back to *moral imagination*, one of the components of the ethical literacy model we articulated in earlier chapters. Scholars in ethical literacy have argued this point. Nancy Tuana, for example, argues that moral imagination "includes empathy for the feelings and desires of others" (2007, p. 375) among a range of other abilities. Henriikka Clarkeburn, in thinking about ethical sensitivity, argues that moral imagination "requires one to 'see' something that is not real in a sense that we could touch or feel it, but something that is real in our minds and within our social existence" (2002, p. 440). The basic idea is that moral imagination requires empathy, ethical literacy requires moral imagination, and understanding digital ethics depends on becoming ethically literate: Empathy is necessary but not sufficient for digital ethical literacy.

So, the next question is: Just what do we mean when we use the term "empathy"?

Defining Key Terms: Empathy, Sympathy, and Other Strange Beasts

"Empathy," as a term, is often interchanged with "sympathy," yet the distinction between the two is regularly ignored. We find this strange since, to us, this dis-tinction is substantial. More importantly here, on our view, keeping those terms distinct is particularly helpful in thinking through why the bigger conversation about empathy and desensitization matters in the digital context.

So when people use the term "empathy," they seem to mean by it some sort of connection to another person. Being able to "walk a mile in another's shoes" is taken to be important if not necessary for ethical decision making. Imagine, for example, that we are trying to understand why someone might think that it is acceptable to speak falsely from a position of power. Likely, from our perspective, this is ethically wrong: Whether you are a Kantian or a utilitarian or a virtue ethicist, lying (especially from a position of power where negative consequences may be more severe) is bad and there are good arguments to support this view. But, if we were able to take this other person's perspective, perhaps we would better be able to understand their own argument (or we would find that they have no argument and are acting from habit or from poor moral character). Moreover, this use of "empathy" enhances our moral sensitivity to a more fundamental point; namely, that others *have* perspectives. This line of reasoning has been used to advance arguments applied to issues ranging from immigration to social justice to animal sentience. More recently, this line has been used to help us consider what it would be like if (or when) advanced artificial intelligence systems have perspectives of their own.

The argument we are making here is that conversations about perspective-taking are important to ethics due to their role in the process of making ethical decisions. In the next section, we will see how this has been more clearly reflected in contemporary ethics literature.

The problem with this generic use of "empathy" is that it obscures the difference between empathy and sympathy. Simply put, sympathy is the capacity of moral agents to feel alongside another. "Moral agents" is a more or less technical philosophical term for those individuals capable of making judgements about matters of right and wrong/good and bad, and acting on those decisions. This set of individuals is under continual revision, but includes at least normally functioning human beings (non-controversially) and perhaps some subset of nonhuman animals (slightly more controversially). Imagine, for example, how you can tell when one of your friends is feeling upset. There are (at least) three reactions you might have. First, you might experience that feeling of *schadenfreude*: that unique German term for taking pleasure in someone else's misfortune. Second, you might experience nothing at all. This sort of disconnection is perhaps rare and notable both as an effect of certain neurobiological conditions or a psychological condition of sociopathy. Third, you might feel upset *on your friend's behalf.* The idea here is that you recognize physical expressions of an emotion in your friend that you, yourself, have previously experienced—and thus can *sympathize* with your friend. For a robust treatment of sympathy in the history of philosophy, take a look at the Internet Encyclopedia of Philosophy entry on the topic (Agosta, 2018).

Empathy, on the other hand, is the condition of *taking on the emotive experiences of the other* yourself. This is, of course, an impossible feat, and the heart of the famous philosophical problem of other minds. That problem suggests that, no matter your epistemological starting point, there is no way to know not only the *content* of other minds but, even more fundamentally, whether or not other minds even exist. Some philosophers have taken what amounts to an ontological shortcut around this problem, arguing that others are fundamental to what it is to be a self in the first place—that you do not exist apart from your relationships with others. Taxonomies of philosophical views on empathy have claimed that "[t]here are at least eight different major variations on neuropsychological theories of empathy that are not equivalent ... though they all tend to position empathy as either the literal experience of another's internal, affective states, or an innate, nonconscious understanding of another at a neurocognitive level, hence differentiating empathy from sympathy or compassion" (Bollmer, 2017, p. 64). Other taxonomies or overviews have argued for two major types of empathy: cognitive and affective (Maibom, 2018). Cognitive empathy "denotes the ability to ascribe mental states to others, such as beliefs, intentions, or emotions," whereas affective empathy "essentially involves affect on the part of the empathizer" (Maibom, 2018, p. 1), which is to say that the empathetic person *feels* rather than *understands* the perspective of the other person.

Yet philosophers' arguments do not necessarily agree, nor do they necessarily set the tone of conversations around such concepts. As a historical example, modernist empiricist David Hume developed a related view he called sympathy, arguing that this human capacity was connected to altruistic behavior. Interestingly, many scholars now take Hume to be talking about empathy rather than sympathy—an important demonstration about the pragmatic ways these terms have shifted over time. And we recognize that this philosophical definition is a specific usage of the term "empathy" and that many use it as a descriptor for other concepts like perspective-taking. This diversity of usage is due, in part, to the enthusiasm around the discovery of biological devices, like mirror neurons, that help govern the ways in which we human animals connect to one another. Social scientists, like psychologists, are especially excited about such discoveries because they take them to lend credence to the foundations of their own research methods and assumptions. In a 2005 cover story for the American Psychological Association's *Monitor*, a staff writer points to scientific literature that suggested these neural mechanisms that allow us to "immediately and instinctively" (Winerman, 2005) understand that others' thoughts, feelings, and intentions are "involuntary and automatic" (Winerman, 2005, p. 48)—they do not rely on any sort of post-experiential rationalizing. But even in thinking about what was then early scientific evidence of neuroscientific explanations for how we believe in others' minds, the author conflates sympathy and empathy. We "recoil in *sympathy*," the author writes, when we see someone get hit with a Frisbee, and notes that the mirror neurons "could help explain how and why we 'read' other people's minds and feel *empathy* for them" (Winerman, 2005, p. 48). The emphasis here is ours, in order to point out one example of ways sympathy and empathy continue to be conflated.

Mirror neurons were first discovered by implanting electrodes in the brains of macaque monkey research subjects. And, even as late as 2016, only a single study had successfully identified human neurons with properties similar to those mirror neurons in macaques: not particularly compelling evidence. Indeed, "the pendulum of scientific opinion has begun to swing towards the skeptics" (Taylor, 2016). Yet additional work from both the social sciences and the neurological sciences has continued to utilize the concept of mirror-neurons, or their loose analogs in human neurology, to help explain human capacities and deficiencies (e.g., Jeon & Lee, 2018). Such work, done in spite of empirical skepticism, suggests that the human capacity to connect to others (whatever we choose to call it) is not only of scientific importance, but also of *ethical* importance.

So What for Digital Ethics?

Empathy in the context of the digital is importantly different from empathy in "traditional" contexts at least insofar as digital environments or technologies *mediate* the relationship between individual human participant users. The conversations around digital storytelling reflect this mediation well. Digital storytelling has been used in the context of engineering ethics, for example, to build what researchers have described as "empathic perspective-taking" (Hess, Beever, Strobel, & Brightman, 2017). Some argue that digital storytelling opens space for importantly novel perspectives and connections, while others level the critique that the sentimentality often utilized in digital storytelling is manipulative and artificial (McWilliam & Bickle, 2017, p. 77). User identities themselves are formed, or packaged, by storytelling. Gardner and Davis (2013) argue that the concept of the "app" is a kind of metaphor for this packaging that has the "consequence of minimizing a focus on inner life, on personal conflicts and struggles, on quiet reflection and personal planning" while at the same time a "broadening of acceptable identities" (Gardner & Davis, 2013, p. 61). An ongoing line of concern related to empathy in the digital has been that digital technologies allow the expression of an "online disinhibition effect," exemplified by evidence that "some individuals may exhibit unusual acts of compassion in online settings, while others may devolve into sarcasm, harsh language, uncouth criticisms, and even cyber bullying" (Terry & Cain, 2016, p. 2). In general terms we have compelling evidence to think that our representations of ourselves through digital technologies and platforms enables us to be different people—or the people we really are—for better, or for worse.

In a different context but along similar lines, the digitization of healthcare has opened new conversations about the role of empathy in health-related domains. Bioethicists know, for example, that developing perspective-taking skills or empathetic connections to patients leads to better health outcomes (Wilkinson, Whittington, Perry, & Earnes, 2017). Others have pointed out the importance of connecting with patients—seeing through their eyes not only creates an empathetic connection between physician and patient but also reduces caregiver stress by helping them better understand patient needs (Graham, 2017; Terry & Cain, 2016). And this same relationship between empathy and conditions of caregiver burnout (namely that better empathy reflects lower rates of burnout) has been evidenced longitudinally and across caregiver types (Wilkinson et al., 2017; Yuguero et al., 2017). However, medical education has not promoted the development of empathy (Terry & Cain, 2016; Wilkinson et al., 2017). Furthermore, on some reads (see Terry & Cain, 2016), the digitization of healthcare means an entirely new landscape of digital empathy that likewise requires attention. As digital technologies like virtual reality training platforms continue to be become less expensive and more ubiquitous across hospitals and care facilities, researchers are able to let caregivers and first-responders have a window into conditions like schizophrenia, PTSD, Alzheimer's, and macular degeneration (to name a few relatively recent success stories) (Terry & Cain, 2016). The sort of window such technologies provides only offers a perspective on part of the experience of the other; namely, into the neurophysiological sensory differences brought about by specific diseases. This form of digital mediation has led some researchers to begin to develop a specific view of empathy as *digital empathy* which they define as the "traditional empathic characteristics such as concern and caring for others expressed through computer-mediated communications" (Terry & Cain, 2016, p. 1). Thus, digital technologies either create or deconstruct empathy—and we do not have sufficient evidence which direction this relationship works for any particular digital technology.

So, empathy is necessary for ethical literacy but seems to face importantly novel challenges in the digital, from the mediation between individuals that occurs in digital environments and through digital technologies. Thus, if conversations about empathy are important for ethics generally, they are even more important for digital ethics.

Digital Desensitization

Since ethics sensitivity is one of the three components of digital ethical literacy, *desensitization* is an important question for digital ethics. By "desensitization" here we mean to focus on the undoing of empathetic skill development that one might have otherwise obtained in analog environments. Whatever you think of the idea of "digital natives" (members of those generations who have grown up alongside the ubiquitous rise of digital media), it seems noncontroversial to point out with others that "the specific features of digital media have seen their dissemination and uptake by youth proceed at an unprecedented pace" (Bennett & Robards, 2014, p. 1). This uptake has resulted in the blurring of boundaries between analog and digital interactions for these individuals particularly and for the rest of us more generally. Carrie James, in her work on this question of desensitization (or, as she calls it, "disconnection"), argues that while empathy is certainly important for interpersonal social-ethical issues like bullying (or cyberbullying), it is insufficient for the complex contemporary ethical issues in which, she writes, there is no one identifiable victim with whom to empathize (2014). While "digital technologies have paved the way for newer, faster, and arguably richer ways to share content and connect with one another than we had at our disposal even just 20 years ago," their implications are just "widely conjectured and beginning to be understood" (James, 2014, p. 11). Thus the question of desensitization is not a zero-sum game: The effects of the digital (whatever those are) are blurred along with the effects of the analog (whatever those are).

Yet, there have been numerous studies done to begin to build links between some aspects of digital information flows and particular sorts of desensitization. A 2011 study done by psychologists found that violent media content desensitizes participants to future content of the same type in ways that sad or funny content does not (Krahé et al., 2011). In this study, researchers utilized biometric data in the form of skin conductance levels and qualitative measures in the form of "a lexical decision task" (Krahé et al., 2011) to measure desensitization, as they define it.

That is an important point: This psychological study took desensitization to be physically and cognitively affective—causing changes to behaviors. This is importantly different from the definition we offered at the beginning of this section. Our focus here is particular; indeed, "[d]esensitization is typically defined as a long-term process in which the repeated or continuous exposure to media violence leads to a decrease in physiological, affective, or cognitive reactions to violence" (Breuer, Scharkow, & Quandt, 2014). Such a generalist definition is meant to cover a full range of changes in reaction or attitude. Just like with all the concepts we have addressed in this book so far, paying close attention to how concepts are defined and used is essential to making good arguments about their ethical implications. Breuer and colleagues thoughtfully point out that not all definitions of desensitization are easily equated: "'Physiological desensitization' cannot be easily equated with emotional or cognitive desensitization, as it might, for example, also indicate boredom or a lack of interest" (Breuer et al., 2014, p. 178). They also point out, as we did in our chapter on embeddedness, that there is little connection between fictional desensitization (like the kind one might receive from violent videogames or movies) and decreases in empathy toward victims of real-life violence (Breuer et al., 2014; Carnagey et al., 2006). And, furthermore, the *type* of engagement seems to matter, too. In their own study, Breuer and colleagues found evidence to suggest that "whether a digital game is being played or just watched can affect the perception and evaluation of violent content" (2014, p. 184). Whether perception and evaluation of content is connected to empathy or perspective-taking is a related but distinct question.

Nonhuman Empathy

Empathy, like agency, is a concept regularly taken as a capacity of only human animals. And even then, as we discussed briefly above, while *only* human animals are considered to have the capacity for empathy, not even *all* human animals are so considered. This idea of anthropocentric privilege has been widely critiqued over the last 30 or so years, with more and more evidence supporting the idea that some other animals, too, can be empathetic. There are some (now) obvious additions to our list of empathetic organisms, including great apes, dolphins, and elephants. But there's scientific evidence to support the idea that, say, voles demonstrate interspecies empathy (defined in this case in a neurochemical and behavioral response sort of way) (Burkett et al., 2016). And while it is good to have empirical evidence to support any hypothesis, it will likely strike you as intuitively plausible that lots of other animals—if not most animals—are capable of empathy. Perhaps it is less plausible to you that nonhuman animals demonstrate the sort of high-level cognitive empathy of which you (we assume) are capable – the sort that allows you to understand what another might be feeling. But, we hypothesize that it is much more plausible to you that many nonhuman animals demonstrate affective empathy; that is, they are capable of feeling what another is feeling. There are lots of conceptual games we might play here to get more nuanced: Are we talking just intraspecies empathy, or *inter*species? Upon just which definition of empathy do our intuitions rest? But the question is an important one to ask. And so, the question for digital ethics is: What are the limits of empathy *beyond* the analog/organismal? If empathy is something that animals can have in various way to various degrees, can the same not be true of digital computational systems? Can such a system have empathy? Can we empathize with one? Should we?

To get at some of these questions in a more practical way, consider the case of Tay, the artificial intelligence inforg, or informationally embodied organism. When Tay came online in late March of 2016, Microsoft touted her as an important experiment in artificially intelligent learning systems. Tay was built to learn from her interactions online with human agents through dialogues mediated by Twitter. Tay was online for less than a day and, after only 16 hours, had been exploited by a series of Twitter users who "taught" her a host of inflammatory and derogatory statements which she began to utilize regularly. Her online persona became such a cultural liability to Microsoft that they humanely euthanized Tay less than 24 hours after her first public interactions (Horton, 2016).

This case is instructive for digital ethics. While there have been analogs throughout history—think simply of any instance where a human child has learned behaviors and phrases that are embarrassing for parents—the speed and scope of Tay's development from naïve well-intentioned child to rough-and-ready racist is unprecedented. This matters when it comes to questions of empathy in the digital. Human development is an analog matter, comparatively slow and staged. Many moral psychologists still hold onto staged or developmental models of ethical decision making; the processes of reasoning are still privileged, and the conditions on which ethical problems are based have enough mass as to change momentum slowly. But in the digital, artificial intelligence systems like

Tay develop at an entirely different pace, decisions are made at lightspeed algo-rithmically rather than psychologically, and the conditions of problems are under constant reorganization. These differences between analog and digital develop-ment fundamentally change the ways we connect and the ways we human agents empathize. Mediated by technology, communicating with a technology, and engaging collectively and simultaneously, human agents put Tay under novel conditions of empathetic relations—and her outcome was a direct result of that digital interfacing.

There are similar examples you might consider here, too. Some of the earli-est artificial intelligence systems pushed up against Turing test limits by con-vincing human interlocutors that they were, in fact, human agents. Online bots, including chat-bots, have ever-increasingly convincing dialogues with interlocutors without giving away that they are artificial and digital. Tech ana-lyst firm "Are You a Human" estimates that 59 percent of online traffic, from gaming to dating sites and social media communications, was driven by bots in 2016 (Kushner, 2016). Users seem to relate to what are functionally home-automation systems like Amazon's Alexa, Google's Home, or Apple's Siri, each of whom can respond to jokes, hold basic conversation, and respond to name cues. These personified artificial systems enable users to search, connect, play, listen, engage, and control by engaging in regular dialogue with the system. The success of these systems is in part due to how familiar they are to engage: like requesting something of a partner, friend, or employee, we simply ask and the digital persona acts. They are programmed to simulate human agents, who we take to be empathetic. Does such technological sleight of hand entail that humans and such artificial systems are able to empathize, or consider one another's perspectives? Or does it give us reason to think that we no longer need each other—or need to connect to one another—as we did in analog conditions in order to fulfil some set of same basic human needs? Questions like these push the ethical necessity of empathy beyond the human and into the technological.

The relationships we develop with such systems have emergent ethical impli-cations that we have never had to consider. In December of 2016, for example, Amazon's Alexa made headlines because of what she might have heard: She might have captured evidence of a murder in Arkansas (Chavez, 2017). Courts had to decide whether or not they could admit Alexa's "testimony" into evi-dence. In this case, Amazon initially denied the request, citing privacy practices, but then relented after the defendant gave specific permission for those record-ings to be released. This sort of digital witnessing of crimes is novel: For a human witness, lawyers could argue that they had an ethical responsibility to offer that evidence in court. This ethical responsibility is a result of their moral agency and, if our argument is correct, moral agency depends on a capacity to empathize (as a necessary condition). But we do not think Alexa has an analogous ethical responsibility, because Alexa cannot empathize—can she?

Empathy Machines

Virtual reality machines, in particular when the technology was young, were billed as "empathy machines." The idea was that virtual reality allowed developers to place users *into* the experiences of others—experiences that they themselves would not have had without the technology. In this way, the story went, we could empathize with others via that technological conduit to their experience. Chris Milk, in a 2015 TED talk, describes this as the true power of virtual reality: "It connects humans to other humans in a profound way that I've never seen before in any other form of media. And it can change people's perception of each other... it's a machine, but through this machine we become more compassionate, we become more empathetic, and we become more connected" (Milk, 2015). More recently, the tone of the conversation around the potential of VR has shifted, as more and more critics point out that claims about VR's empathy-building potential are conceptually confused and lacking in empirical support (Robertson, 2017). Further, for what perspectives VR might offer, those same perspectives can be manipulated (Robertson, 2017). One critic argues in no uncertain terms that "The rhetoric of the empathy machine asks us to endorse technology without questioning the politics of its construction or who profits from it. Empathy is good, and VR facilitates empathy, so therefore VR is good—no questions please" (Yang, 2017, para. 4).

In this critical view, VR is a technology not by which users empathize but through which users appropriate the experiences of the other. Paul Bloom, author of *Against Empathy*, gave a powerful example of the ethical implications of such appropriation. "In reality," he notes,

> Empathy can go both ways. Somebody could develop a machine giving empathy for Syrian refugees, but also a machine giving empathy for someone who lost his job because a Syrian refugee took it and we watch him standing hungry in a food line. Or, to use one of Trump's favorite examples, imagine a machine giving you empathetic feelings for someone who has been assaulted by an undocumented immigrant. Sure, empathy will support my side, but it's not trivial to imagine how empathy could be exploited for any side.
>
> *(Chen, 2016, para. 5)*

The debate around VR and empathy indicates that developers want their digital technologies to help human beings better connect, in this case to the experiences of intraspecies members. Developers recognize the importance of empathy for making ethical decisions, and are split about the role that digital technologies might have in advancing this end. The following case study helps get at questions about the targets of empathy in digital ethics, which include all but not only human others.

CASE STUDY: GAMES AND DESENSITIZATION

On August 26, 2018, a 24-year-old Baltimore man shot 12 victims, killing three including himself, during participation in a Madden 2019 videogame tournament in Jacksonville, Florida (WLWT5 National Desk Staff, 2018). You are aware of the debate circling around the relationship between videogames and violence. So you perform a quick Google search for desensitization in videogames and find that the top link is an April 9, 2016 *Psychology Today* article titled "Violent Video Games Can Trigger Emotional Densensitization" (Bergland, 2016). The second link is a May 8, 2017 *WebMD* article titled "Violent Video Games May Not 'Desensitize' Players" (Dotinga, 2017).

Given this evidence, how do you go about evaluating the relationship between the context of digital gaming and desensitization?

Next Up

In this chapter, we have argued that empathy is a necessary condition of ethical decision making. We have pointed out conceptual issues, like the conflation of empathy and sympathy, and pointed toward some ethically significant implications of such confusion. We have considered under what conditions desensitization might occur around or through digital technologies and offered several cases and thought experiments as frames through which to think critically about this concept in digital ethics. In the next chapter, we review a lower level of interaction—the individual—and consider the nature of agency related to digital ethics.

References

Agosta, L. (2018). Empathy and sympathy in ethics. *Internet Encyclopedia of Philosophy.* Retrieved from https://www.iep.utm.edu/emp-symp/#SH4a.

Associated Press. (2018). 2 12-year-olds arrested for cyber bullying in connection with suicide of 12-year-old girl. *ABC Eyewitness News.* Retrieved from https://abc13.com/2-arrested-for-cyber-bullying-after-12-year-old-girls-suicide/2983420/.

Bollmer, G. (2017). Empathy machines. *Media international Australia, 165*(1), 63–76.

Bennett, A., & Robards, B. (Eds.). (2014). *Mediated youth cultures: The Internet, belonging and new cultural configurations.* Basingstoke, UK: Palgrave Macmillan.

Bergland, C. (2016). Violent video games can trigger emotional desensitization. *Psychology Today.* Retrieved from https://www.psychologytoday.com/us/blog/the-athletes-way/201604/violent-video-games-can-trigger-emotional-desensitization.

Breuer, J., Scharkow, M., & Quandt, T. (2014). Perception and evaluation of violence in games. *Journal of Media Psychology, 26*(4), 176–188.

Burkett, J.P., Andari, E., Johnson, Z.V., Curry, D.C., de Waal, F.B.M., & Young, L.J. (2016). Oxytocin-dependent consolation behavior in rodents. *Science, 351*(6271), 375–378.

Carnagey, N.L., Anderson, C.A., & Bushman, B.J. (2006). The effect of videogame violence on physiological desensitization to real-life violence. *Journal of Experimental Social Psychology, 43*, 489–496.

Chavez, N. (2017). Arkansas drops murder charge in Amazon Echo case. *CNN*. Retrieved from https://www.cnn.com/2017/11/30/us/amazon-echo-arkansas-murder-case-dismissed/index.html.

Chen, A. (2016). Paul Bloom on why VR empathy projects won't save the world. *The Verge*. Retrieved from https://www.theverge.com/2016/12/6/13857268/paul-bloom-psychology-against-empathy-virtual-reality-politics.

Clarkeburn, H. (2002). A test for ethical sensitivity in science. *Journal of Moral Education, 31*(4), 440–441.

Ditch the Label. (2017). The annual bullying survey 2017. Retrieved from https://www.ditchthelabel.org/wp-content/uploads/2017/07/The-Annual-Bullying-Survey-2017-1.pdf.

Dotinga, R. (2017). Violent videogames may not 'desensitize' players. *WebMD*. Retrieved from https://www.webmd.com/balance/news/20170308/violent-video-games-may-not-desensitize-players-brain-scans-suggest#1.

Florida Legislature. (2018). The 2018 Florida statutes. *Online Sunshine*. Retrieved from http://www.leg.state.fl.us/statutes/index.cfm?App_mode=Display_Statute&URL=1000-1099/1006/Sections/1006.147.html.

Garcia-Perez, R., Santos-Delgado, J-M., & Buzon-Garcia, O. (2016). Virtual empathy as digital competence in education 3.0. *International Journal of Educational Technology in Higher Education, 13*, 30.

Gardner, H., & Davis, K. (2013). *The app generation: How today's youth navigate identity, intimacy, and imagination in a digital world*. New Haven, CT: Yale University Press.

Gordon, E., & Schirra, S. (2011). Playing with empathy: Digital role-playing games in public meetings. *C&T' 11 Proceedings of the 5th International Conference on Communities and Technologies*, 179–185.

Graham, M. (2017). The empathy effect: How virtual reality cuts caregiver burnout. *Dell Technologies Perspectives*. Retrieved from https://www.delltechnologies.com/en-us/perspectives/the-empathy-effect-how-virtual-reality-cuts-caregiver-burnout/.

Hess, J.L., Beever, J., Strobel, J., & Brightman, A.O. (2017). Empathic perspective-taking and ethical decision-making in engineering ethics education In B. Newberry, D. Michelfelder, & Q. Zhu (Eds.), *Philosophy and engineering: Exploring boundaries, expanding connections* (pp. 163–179), Cham, Switzerland: Springer.

Horton, H. (2016). Microsoft deletes 'teen girl AI after it became a Hitler-loving sex robot within 24 hours. *The Telegraph*. Retrieved from https://www.telegraph.co.uk/technology/2016/03/24/microsofts-teen-girl-ai-turns-into-a-hitler-loving-sex-robot-wit/.

James, C. (2014). *Disconnected: Youth, new media, and the ethics gap*. Cambridge, MA: MIT Press.

Jeon, H., & Lee, S-H. (2018). From neurons to social beings: Short review of the mirror neuron system research and Its socio-psychological and psychiatric implications. *Clinical Psychopharmacological Neuroscience, 16*(1), 18–31.

John, A., Glendenning, A.C., Marchant, A., Montgomery, P., Stewart, A., Wood, S., Lloyd, K, & Hawton, K. (2018). Self-harm, suicidal behaviours, and cyberbullying in children and young people: Systematic review. *Journal of Medical Internet Research, 20*(4), e129.

Kushner, D. (2016). Scammers and spammers: Inside online dating's sex bot con job. *Rolling Stone*. Retrieved from https://www.rollingstone.com/culture/culture-news/scammers-and-spammers-inside-online-datings-sex-bot-con-job-189657/.

Krahé, B., Möller, I., Huesmann, L.R., Kirwil, L., Felber, J., & Berger, A. (2011). Desensitization to media violence: Links with habitual media violence exposure, aggressive cognitions, and aggressive behavior. *Journal of Personality and Social Psychology, 100*(4), 630–646.

Maibom, H. (Ed.). (2018). *The Routledge handbook of philosophy of empathy*. New York, NY: Routledge.

McWilliam, K., & Bickle, S. (2017). Digital storytelling and the 'problem' of sentimentality. *Media International Australia, 165*(1), 77–89.

Milk, C. (2015). How virtual reality can create the ultimate empathy machine. *TED*. Retrieved from https://www.ted.com/talks/chris_milk_how:virtual_reality_can_create_the_ultimate_empathy_machine/transcript#t-613612.

Robertson, A. (2017). VR was sold as an 'empathy machine' – but some artists are getting sick of it. *The Verge*. Retrieved from https://www.theverge.com/2017/5/3/15524404/tribeca-film-festival-2017-vr-empathy-machine-backlash.

Taylor, J. (2016). Mirror neurons after a quarter century: New light, new cracks. Retrieved from http://sitn.hms.harvard.edu/flash/2016/mirror-neurons-quarter-century-new-light-new-cracks/.

Terry, C., & Cain, J. (2016). The emerging issue of digital empathy. *American Journal of Pharmaceutical Education, 80*(4), Article 58.

Tuana, N. (2007). Conceptualizing moral literacy. *Journal of Educational Administration, 45*(4), 364–378.

UBC News. (2012). Cyberbullying and bullying are not the same: UBC research. *UBC News*. Retrieved from http://news.ubc.ca/2012/04/13/cyberbullying-and-bullying-are-not-the-same-ubc-research/.

Yang, R. (2017). 'If you want in someone else's shoes, then you've taken their shoes': empathy machines as appropriation machines. *Radiator Design Blog*. Retrieved from https://www.blog.radiator.debacle.us/2017/04/if-you-walk-in-someone-elses-shoes-then.html.

Yuguero, O., Forné, C., Esquerda, M., Pifarré, J., Abadías, M. J., & Viñas, J. (2017). Empathy and burnout of emergency professionals of a health region: A cross-sectional study. *Medicine, 96*(37), e8030.

Wilkinson, H., Whittington, R., Perry, L., & Earnes, C. (2017). Examining the relationship between burnout and empathy in healthcare professionals: A systematic review. *Burnout Research, 6*, 18–29.

Winerman, L. (2005). The mind's mirror. *Monitor on Psychology 36*(9), 48.

WLWT5 National Desk Staff. (2018). 3 Dead, including gunman, after shooting at Florida videogame tournament. *WLWT5*. Retrieved from https://www.wlwt.com/article/jacksonville-florida-video-game-shooting/22833779.

8

AGENCY, AUTONOMY, AND ACTION

In this chapter we define agency in two ways—one operationally, as we use the term in everyday conversation, and one theoretically and philosophically, allowing us to think about the deeper long-term ethical implications of agency and digital technology. We next discuss how agency changes in the digital age—both how our notions of control and power are reconfigured when considering the digital and how the characteristics and affordances of digital media complicate our relationships to information and audience. In order to further investigate these themes, we then discuss two specific aspects of re-situated agency—time and sociality—and explain how temporal and social perceptions are shifted by our ways of using and thinking with digital tools. We consider the impacts these aspects have on the relationship between agency and autonomy, the capacity to act freely and intentionally. Our experiences using these tools in our everyday lives can complicate our ability to socialize in physical spaces and can open conversations about digital addiction, both topics we discuss in this chapter.

CASE STUDY: "TAY-KEN" OVER

Questions of agency in the digital appear regularly, even if our normal discourse does not recognize them as such. Recall Tay, the artificial intelligence system that Microsoft brought online in March of 2016. Tay was designed to simulate human conversation online, in a public experiment to pass the Turing test, a test named for philosopher Alan Turing wherein an artificial conversational system would fail to be distinguishable from a human speaker. Tay's success quickly became a paradigmatic digital ethics case, as users began to twist and train her into a simulation of a morally vicious racist, sexist, and abusive online persona. It only took a day. Around 8:00pm on the 23rd, Tay was responding to chats with statements like "can I just say that im stoked to meet u? humans are super cool." By 11:45am on the 24th, she was responding with statements like, "Hitler was right I hate the jews" (Vincent, 2016). Tay was influenced, in ways analogous to how a human person online might be influenced but at much greater speed and scope, by online communities—and with that influence came important questions about whether and to what extent she had agency and, perhaps more importantly, how autonomous that agency was. Indeed, it is not hard to imagine an AI like Tay that is completely indistinguishable from a human person: Many such artificial systems interact with us online daily right now. And it is also not hard to imagine a next step, wherein a conversational AI system would become embodied. Boston Robotics' mechanical creations have advanced in just a few short years from waddling awkwardly to opening doors and jumping uncanny valleys in a single bound. And if it talks like a person, and walks like a person, it's a person, isn't it?

Autonomy and the Autonomous

Concepts like autonomy and agency are immensely complicated, yet used in our everyday language as if they were ever-so-simple. And there is risk here: Both are central to ethical action, so to oversimplify these concepts runs the risk of oversimplifying their ethical implications. Autonomy is the freedom to act, while agency is the capacity to act. If we feel as though we have autonomy when a decision must be made, for example, then we feel like what we say and do has the potential to influence the outcome of that decision in a meaningful way. When we have agency in these decisions, we feel as though our actions and behaviors have impact on our environments. Agency is a powerful construct of human existence because it makes us feel as though we have a say in our own destinies. Our actions, when imbued with agency and empowered by autonomy, have meaning and are important in the context of our environments and the individuals and objects with which we interact in those environments.

Agency can also be described more formally as intentional individual action, or as mental, shared, collective, relational, or artificial (Schlosser, 2015), depending on the theoretical context. We can further differentiate simple agency (the capacity to act) from autonomy, the capacity to act freely and intentionally. The two are intimately related: You cannot have autonomy without agency. The latter concept, autonomy, is deeply connected to ethical concerns about liberty, security, and justice. But, like agency, autonomy is a richly complex concept—*so* complicated, says philosopher Richard Dworkin, that the only constant features among various definitions are that it is a feature of persons and that it is a desirable quality to have

(1988, p. 6). Not a high bar. We will continue to talk in terms of agency but, like Schlosser's definition above, our focus on ethical implications necessarily compel us to think of agency in terms of its partnership with autonomy: We are taking it for granted that agency is an active force within the digital, and that the interesting problems are at its intersection with *free and intentional* action.

One key conception of agency for us comes from Emirbayer and Mische (1998), who use what they describe as a "chordal triad of agency" (p. 970) to characterize agency as "the temporally constructed engagement by actors of different structural environments—the temporal-relational contexts of action—which, through the interplay of habit, imagination, and judgment, both reproduces and transforms those structures in interactive response to the problems posed by changing historical situations" (p. 970). While more complex than our casual understandings of agency as linked to power and control, this model opens up possibilities for thinking about agency not only in the present, but also in the past and in the future. This broader theoretical articulation of agency is particularly well suited to thinking about the ethics of digital machines and the accompanying re-situation of agency such machines may catalyze.

One critical component of Emirbayer and Mische's expanded description is found in their notion of a "temporally constructed engagement," which highlights the primacy of time in their conceptualization of agency. In our operational definition, time is related to being "in the moment," of knowing that the actions or decisions we make now will have some significant bearing on the outcome of a situation. However, Emirbayer and Mische do not just consider agency in relation to the present, but also in relation to the past and the future. They describe "a temporally embedded process of social engagement, informed by the past (in its habitual aspect), but also oriented toward the future (as a capacity to imagine alternative possibilities) and the present (as a capacity to contextualize past habits and future projects within the contingencies of the moment)" (p. 963). Accordingly, when we act with agency, we are not only feeling powerful and in control of the moment at hand, but we are also considering our past actions and the impact of our past and present decisions on future outcomes. Such considerations rely upon a belief that we have free will to construct our own paths through life. We can think about agency not only as the ability to influence the present, then, but also as the ability to see how our prior decisions and future possibilities all interact to form the conditions of our everyday consciousness and decision making. In this way, agency can be conceptualized as a human quality closely related to our experience of time in which we are moral beings who act with free will to shape the futures we desire for ourselves and others. The results of these actions provide evidence of our values and thinking.

Our digital devices rely upon time in more fundamental ways. After all, many digital technologies like the modern computer depend on an internal clock for basic functionality, and other gadgets like alarm clocks and mobile phones all have various programs and applications for measuring and working with time in quantifiable units (e.g., alarm clocks, stop watches, countdown timers, calendars, etc.). However, what Emirbayer and Mische are discussing are human and subjective notions of time and our conscious relationship to time. While traditional computers repeat actions over time, they are generally not thought of as forming habits. In humans, habits can respectively be broken or reinforced, while computers would need new programming to change repeated actions over time. Similarly, when Emirbayer and Mische are discussing imagination, they reference a human trait that is not generally associated with computing technologies. Judgment, too, is a human characteristic, although we have been discussing technologies that may need to make decisions that look very similar to the judgments humans make. Two examples are the autonomous vehicle that must decide between imperfect outcomes when a traffic accident is imminent and the companion robot that must decide how to deal with a human owner who will not take his medication. And artificial intelligence systems like Tay and those governing self-driving vehicles continue to develop in scope and sophistication, pushing against this idea that all and only *human* agents can form habits, imagine new situations, and make (ethical) judgments.

Another notable aspect of Emirbayer and Mische's (1998) definition is that they have chosen to theorize agency in the plural, as indicated in their use of the term actors, rather than actor, when they write about "the temporally constructed engagement by actors of different structural environments" (p. 970). Sociality is another important aspect of agency, since often our manifestations of control and power are dealing with not only objects in our environment, but also other people. These notions of power and control are complicated by what happens to these ideas when they affect different numbers of individuals. For example, one individual's sense of control may be exaggerated or diminished when that individual is associated with large groups of individuals with similar or opposing viewpoints. Further, the emotional dynamics of social groups are also critically important when considering power and control; after all, affective (emotional) rhetorical techniques are used strategically in important social power activities such as politics. Here, political discourse often relies upon strong emotional reactions in audiences to build support for legislative change.

As these casual and formal definitions show, agency is a complex cognitive, social, and emotional construct that is of vital importance in a human's ability to feel self-sufficient and empowered. In the next section of this chapter, we more broadly discuss some of the implications for agency in the digital age. We will then drill down more specifically into two of the areas Emirbayer and Mische's definition suggests are important for a more robust conceptualization of agency: time and social relationships.

From Apple Inc. to Teenage Repellant: Agency in the Digital Age

In both our colloquial definition and the extended theoretical definition from Emirbayer and Mische, autonomous agency seems essential to humans, but not necessarily to machines. After all, what do machines care about such things as control, power, and social relationships? To be sure, a sufficiently advanced artificial intelligence may care very much about these things, since each can be attributed to intelligence in various ways (see, for example, Luciano Floridi's arguments (2013) about the morality of artificial agents). However, there are more subtle ways in which power, control, social relationships, and agency have been re-situated by our existing digital technologies. Those subtle shifts of autonomy and agency are the primary topics of this chapter.

The digital shifts the status of our feelings of power in digital environments and our sense of control over our digital tools and technologies. Many of us probably feel like we are in control of *certain* aspects of our digital experiences, at least. We can choose what images and words to post on our websites or social media communities, for example, and we can choose what music to download, when to play it, and at what volume. Similarly, we can choose which pictures to take with our digital cameras and we have control over with whom we share our digital images and how we distribute that information to friends, family members, or the general public. We may also choose to incorporate digital works into our professional lives, working with various software programs and digital assets to solve problems in our workplaces and perhaps to produce artistic or entertainment-oriented products for distribution to public or private audiences.

Our notions of agency are complicated, however, by certain aspects of the digital experience and our behaviors in digital domains. First, many individuals who use digital technologies are not what we would call "power users." Instead, they learn how to use the devices well enough to accomplish what they need to accomplish, and leave it at that—they have only basic procedural literacy of digital technologies. However, while the individual may be in at least partial control doing the things she wants to do, there are other things that may be happening "behind the scenes," so to speak, that are outside her sphere of control and understanding. These may be mundane technical operations, like transferring data from long-term to short-term memory storage, or they may be more sensitive operations for privacy and security, such as exchanging information behind the scenes of a web browser. Such behind-the-scenes activity may not be intrusive while the software is being used, but it allows advertising and media companies to build up longitudinal profiles of our content preferences and behaviors (recall the Cambridge Analytica debacle or the microtargeted advertising examples we discussed in Chapter 6). It is unlikely, then, unless one is a technical guru and a digital savant, that one truly has power and control (and, therefore, robust digital literacy) over *all* aspects of their digital experience.

In other cases, power may not lie in the complexity of the technology and the limitations of our knowledge as operators, but rather in the structure of control levied by its designers. Apple Inc. is a case in point. As we discussed in Chapter 6, Apple has demonstrated their philosophy toward third-party content and functionality with what has been described as a "walled garden" approach to technology platforms, beginning with the introduction of the iPod in 2001

(Murphy, 2017). In such environments, hundreds of millions of Apple users are given digital platforms in which to choose content and programs. This environment offers the illusion of choice and control through the sheer number of possibilities, but selections are in fact tightly restricted in order to adhere with strict corporate guidelines. With a central authority (Apple, in this case) that must approve any content additions to a digital repository, we are led back into the same editorial model that predated the Internet Age. If a user wishes to download a particular program on her iPhone, she can select and download curated programs and content that have been approved by the corporation, which are vast in scale and variety. However, rather than true control and power over the computing environment and its resources, the user is experiencing only a subset of content that has been approved by Apple employees and Apple Inc.'s App Store Review Guidelines (Apple Inc., 2017). As Murphy (2017) writes, "Apple handpicks every app that goes live in its store, sometimes rejecting apps for no good reason, or because a foreign power tells them to. It's a system built in the image of its creators, and it's such a beautiful garden that it has paid off handsomely for Apple" (para. 3).

On the other hand, there may be a case to be made for quality control and security and for editorial oversight to ensure Apple's notions of decency are upheld. In Apple's own words, "We will reject apps for any content or behavior that we believe is over the line. What line, you ask? Well, as a Supreme Court Justice once said, 'I'll know it when I see it.' And we think that you will also know it when you cross it" (Apple Inc., 2017, para. 5). The U.S. Supreme Court Justice Apple refers to was Potter Sewart and the reference in question was from his 1964 opinion rendered on Jacobelis v. Ohio. Whether or not this form of editorial control derived from a court case decided over 50 years ago is necessary to uphold the moral values of Apple, it is clear that the use of a nebulous and subjective "line" to determine the appropriateness of content adds further power to Apple's moderation abilities. Lines can be drawn and redrawn over time, adding both flexibility and vagueness to Apple's design guidelines. While Apple's claim is that such careful screening processes ensure quality control and minimize security risks, they are also gaining additional economic and functional advantages by insisting upon this approach to content and software moderation.

Another complicating factor with agency in our digital world is found in the way in which digital content is structured and stored on the web with visibility in mind. One example is the desire for pervasiveness in digital media. Content developers who develop digital assets for the World Wide Web, for example, are increasingly thinking about how to develop "sticky" content that ranks highly in Internet search engines and that encourages web visitors to return and re-engage with content repeatedly (Kominers, 2009). In contrast to the Apple model in which a corporation's employees approve content, sticky content complicates agency because the power of distribution and circulation shifts from a hierarchical, human-centered editorial model to a new model in which the power of visibility and access is transferred to algorithms, such as Google's PageRank algorithm used for website search results (Kamvar, Haveliwala, & Golub, 2004). As designers and developers move from writing content that pleases a human editor to one that must also (or, perhaps, only) please a computer algorithm for search engine rankings, then agency has been at least partly re-situated from the human to the computational domain. Here is a place where differentiating between agency and autonomy is particularly helpful: If agency re-situates from human to computational, what happens to autonomy? Does it too re-situate, so that autonomy is shared between human and non-human systems? Or do human agents *lose* autonomy as they relinquish agency in the digital?

Agency can also be re-situated in the other direction, however, by transferring power in a way that enhances or augments an individual or group's control. For example, consider Donald Trump's use of Twitter, which allowed him when he was running for office to quickly and efficiently disseminate messages and calls to actions to millions of Americans who felt disenfranchised. By using Twitter as a surrogate voice, his campaign built a political platform for mobilizing the working class and introduced a new paradigm for political communication discourse that we had not seen in prior campaigns. This type of power is not limited to the commander in chief, though. Boler (2008) writes of digital media's ability to give everyday citizens the power to intervene and change structures in politics through activities such as establishing alternative media outlets, broadcasting independent podcasts or community radio stations, participating on discussion forums, posting blogs, and becoming citizen journalists by reporting on news from their everyday lives.

Another practical example that highlights the complexity of re-situated digital agency is found in the work of a shopkeeper who developed a "teenager repellent" to prevent loitering in front of his shop. He invented a digital device that played a high frequency sound that was annoying to young ears but inaudible to adults (Buckingham, 2008). In this case, the digital device re-situated agency in a sense by taking over the role of an older adult sitting in a rocker and yelling for kids to stay off his lawn. This achieved the same net effect of reducing teenage loitering, but using a different functional implementation. However, with this re-situation also comes new moral dimensions to be explored. For example, digital technologies do not (yet) possess biological bodies, so they do not tire. What does this mean when they are acting as power-imbued surrogates for humans? Is it morally acceptable for the teenager repellant to be on for 24 hours a day, for example, long after the shopkeeper has returned home to bed and the store has been closed? Additionally, what if a teenager is a legitimate customer looking to make a purchase? While a human could use thoughtful discretion to tell the difference, a digital device such as this is incapable of making such judgement and all customers capable of hearing this range of audio, regardless of intent, will be turned away.

Many of these new digital possibilities also raise ethical questions relating to the agency of larger social groups. For example, what are the acceptable moral parameters for Twitter's use in mobilizing crowds of people toward action?

Unlike traditional journalists, large numbers of individual users with no common code of ethics to follow complicate the traditional paradigms of information reporting and dissemination during major events such as the Arab Spring (Lotan et al., 2011). In addition to the re-situation of agency from humans to computers (as Tweets can coordinate much more quickly and efficiently than door-to-door human networking) we also have a re-situation of agency from traditional news and television journalism to digital citizen activists, or cyberactivists (Eltantawy & Wiest, 2011). These cyberactivists use the power of the Internet and networked devices to mobilize groups for collective action. Whether or not these activities are "good" or "bad" to a particular group or demographic is often a matter of perspective; one can often find equal numbers of demonstrations and counter-demonstrations around controversial subjects such as religion, gun control, and abortion. What is different now is that social media makes the assembly and control of such gatherings easier to manage and information can be more quickly disseminated amongst national and international audiences.

In the next two sections, we will look more closely at two specific manipulations made possible through digital technology. First, we will explore some of the ways in which digital devices affect our social relationships. Here we specifically consider research done with young people and their devices as an example of resituated agency in social contexts. Next, we will explore time, and discuss how computing complicates our chronologies. The chapter will close on this thread by providing a case study and a thought experiment on the topic of electronic memorials to our deceased loved ones.

Virtual Good and Evil in Networked Spaces: Digital Agency and Sociality

As we consider the impact of digital technology on agency, we can ask ourselves how our ability to exert influence is changed in multiple directions. We have already discussed examples in which agency has either been enhanced or diminished by digital technology, but in what other, perhaps less obvious ways, might power and control also be manipulated by our technologies? What does digital technology either newly afford us or take away from us that was previously available? Sherry Turkle (2011) argues that our relationships with technology have left us "alone together" by offering "the illusion of companionship without the demands of friendship" (p. 1). We see similar sentiment in editorials by publications such as the *New York Times* that critique technologies that socially isolate our young people, leverage addictive technologies to make money, and monopolize communications to invade our private lives and "impose unfair conditions on content creators and smaller competitors" (Brooks, 2017, para. 10). We also see problematic social behaviors within specific domains like communications forums, such as the well-documented public sexual harassment on multi-user Internet forums. One infamous example occurred in an early Internet multi-user environment named LambdaMOO in which virtual inhabitants were virtually raped by a user named Mr. Bungle (Dibbell, 1993). Such vile virtual behaviors are due in large part to the anonymity of the medium and the lack of empathy some users have for other individuals in digital settings (and we previously discussed anonymity and empathy in earlier chapters).

Each of these examples speaks to digital technology's manipulation of our control over social relationships. While many new social possibilities are created in digital environments, the social is now directly mediated by the technological, and transfers of power can occur in multiple ways. We have already addressed the possibility of editorial control moving from human to algorithmic control, such as Google's PageRank algorithm for ranking content in search engines or Facebook's live feed algorithm for determining which friends' posts rise to the top of one's news feed. However, as Dibbell's (1993) LambdaMOO example points out, there are also sometimes human transfers of power, as those who are more adept at using technology experience power and control in virtual worlds that they would not have in the physical world. This can lead to rude, abusive, and immoral behaviors in these virtual environments.

On the other hand, there are examples where technology provides access or social communication to physically challenged individuals like Dr. Stephen Hawking (Medeiros, 2015), allowing him to continue speaking, publishing, and sharing scientific ideas long after his native biological capabilities made such activities physically impossible for him (due to ALS). That such activities were done virtually, with an electronically synthesized voice and an intelligent recognition system, meant that a negotiation of agency was necessary. In order to express himself in the way he wished, Dr. Hawking was challenged to learn the intricacies and limitations of the technology, to cede control while also gaining control he had lost. However, the technology itself also adapted to his communication patterns using predictive analytics and contextual menus, providing a level of control that adjusted based on his usage patterns.

Not every example of changing social dynamics is as extreme as the cases of Mr. Bungle and Dr. Hawking suggest, however. Of course it is true that not all relationships are despicable (as we saw with Mr. Bungle's operator's decision to digitally molest) or positively life altering (as we saw with Dr. Hawking's abilities to communicate with a digital voice). In many instances, the shifts are more subtle. For example, in regards to social media, although we are virtually connected to more people than ever before, the work of media critics suggests that this comes at a cost to our real world relationships and our ability to engage with other individuals in "real life," particularly at a young age. These virtual relationships are also mediated by the algorithms that control the social media software, as we discussed in Chapter 7. This means that in applications such as Facebook, the algorithms that control our information feeds determine which messages we see from which friends at any given time, further diminishing our control and power over communication in that environment. These algorithms also shape our exposure to news and can craft ideological bubbles that limit our exposure to alternate viewpoints and beliefs around pressing societal issues (Bakshy, Messing,

& Adamic, 2015). Algorithms can also be manipulated deliberately with malicious intent, as we discussed in Chapter 6 with regards to the attempted suppression of African-American voters.

These algorithms can also persuade us to spend more time online and less time engaging with individuals in the physical world. Internet Addiction Disorder (IAD) affects between 1.5 and 8.2 percent of U.S. and European Internet users (Cash et al., 2012). Patterned behavior classified as addictive implies a loss of control—one *has* to have another drink, or take more drugs, or overeat, or smoke, or gamble on one more horse race, or play another social mobile game, such as *Candy Crush Saga* (Chen & Leung, 2016). In their analysis of mobile gaming addiction, Chen and Leung (2016) explain the broadening of our understanding of addiction, which traditionally focused on the need for a physical substance, to include contemporary behaviors involving technology including surfing the Internet or playing videogames. That these newer forms of technological addiction are mediated by technology provides another level of complication to addictive behaviors. With technology in partial control of the process, we now have precisely manufactured algorithms, reminders, and mechanics that can fine tune and expertly manipulate their users into the patterns of behavior necessary to sustain the virtual experiences over time and make more money for the companies that own them.

Another example of this re-situation of social agency through technology is found in studies about how children interacted with virtual pets beginning in the late 1990s. Returning to Turkle's (2011) work, she writes extensively about the virtual pet Tamagotchi, released in 1997, and interviewed children (including her own daughter) about how they felt about and interacted with these digital toys over time. The device, shaped like an egg, showed a pixelated image of a virtual pet that children were asked to care for in various ways, by pressing buttons and interacting with the device over time. As Lawton (2017) describes, "The 'animism' in the Tamagotchi made users interact with the technology like real pet owners, and come to tolerate the behaviours of the Tamagotchi as they would an untrained pet. The impatience of the Tamagotchi would soon develop tolerance in the user, suddenly enabling a new, more demanding relationship with technology" (p. 3). Turkle's research supported this claim as well; she found that children like her daughter found their Tamagotchis to be "alive enough to care for" (p. 31) and that the simple social robots' dependence on human nurturing and care was seen by their caretakers as making the devices more alive. Nurturance, in Turkle's analysis, "is the 'killer app'" (p. 32). When we spend time "caring for" or "nurturing" our virtual devices or content, then our agency has been resituated. This is because taking care of something can also be seen as exerting power or control over that thing (generally for altruistic rather than devious reasons).

Most striking in Turkle's research is how the children responded when their virtual pets "died." Rather than simply resetting the Tamagotchi device and growing a new pet, the children were reluctant to do this. They preferred to buy new devices and start over from scratch on their new toys. As Turkle explains, "They don't like having a new creature in the same egg where their virtual pet has died" (p. 33). In these situations, rather than taking advantage of the affordances of the digital, such as its ability to be wiped and restarted with a clean program, the children often transferred their ideas about the social observances around death onto the devices. One child interviewed by Turkle, Sally, chose to "bury" her deceased Tamagotchis "with ceremony in her top dresser drawer" (p. 33). In Sally's mind, the digital device's functional use over time was not tied to the lifespan of the technological hardware, but rather to the existence of her virtual pet. Indeed, as the next section demonstrates, time is another factor when considering the re-situation of agency in our digital world.

The Physical Rise and Virtual Fall of Justine Sacco: Digital Agency and Time

There are at least three different ways that agency is re-situated through digital technology due to temporal factors. First, because digital technology is pervasive and networked, the *perceived compression of time* through rapid distribution and redistribution of content allows us to more efficiently and directly exert control over our networks. As our modes of communication have evolved from a purely oral tradition to print texts to the telegraph to electronic communication on the Internet, the speed of transmission has greatly increased the power and influence, good or bad, that one's digital messaging can have on an audience. There are many instances in which an ill-advised Tweet or thoughtless social media post led to severe repercussions for the original author.

One striking example of this time compression is chronicled in the *New York Times Magazine* in their coverage of an incident involving a young woman named Justine Sacco. We mentioned this example briefly in the Introduction, but here we will examine it in more detail. The *Times* article tells the story of Ms. Sacco, a corporate communications director, who was taken to task by Internet communities for her communication using Twitter. She ultimately lost her job due to her insensitive and offensive comments that she thought were appropriate for the medium, in which sarcasm and off-color humor are frequently used for comedic effect. She wrote: "Going to Africa. Hope I don't get AIDS. Just kidding. I'm white!" (Ronson, 2015). Sacco, who at the time was a 30-year-old senior director of communications for a large firm, posted this message to her Twitter account just prior to an 11 hour flight, the last connecting leg to her ultimate destination in Cape Town, South Africa. When the plane landed, she found that the message had "gone viral," trending at #1 on Twitter, and her phone was filled with messages from concerned friends and colleagues who had been monitoring the unfolding saga while Sacco was on her flight (Ronson, 2015).

While in the air and unbeknownst to Sacco, the message had been posted, reposted, and vilified in online communities who were both outraged by the message and intrigued by the spectacle of Sacco being offline and out of the loop while the drama unfolded. As Ronson (2015) described, "The furor over Sacco's tweet had become not just an ideological crusade against her perceived bigotry but also a form of idle entertainment. Her complete ignorance of her predicament for those 11 hours lent the episode both dramatic irony and a pleasing narrative arc" (para. 11). In these types of "viral" situations, it often appears as though the digital message has taken on a "life of its own" and this is made possible through the perceived compression of time. In such scenarios, major communicative movements happen very quickly and with much fervor and online impact. This anthropomorphic metaphor or ontological position (depending on your view) for a digital asset, as something with agency and a "life of its own," is an example of what we mean by the re-situation of agency due to temporal compression. While Sacco was certainly in control of the original message and whether or not to post it (and clearly made a bad decision), she ceded autonomous agency to the digital systems when she lost access to Wi-Fi due to her plane's departure.

One can draw similar parallels to the autonomous vehicles discussed in Chapter 2; since computers are so adept at making fast and accurate decisions in emergency contexts, like unexpected braking scenarios, it has now become commonplace in vehicle adaptive safety features for the technologies to take control from their human operators in order to avoid accidents. This is a second characteristic of time that changes our dynamic to agency—computers can perform many operations much more quickly than humans can. In other words, the *scale of time is different for computers than for humans*. In terms of physics and the mechanics of how time passes, of course, this is not true. But psychologically speaking, it is very true, since thousands of operations may happen in a computer within a second of a human's existence. This requires computing operations to be measured at a very fine and granular degree, using microseconds or even smaller units for measurement.

This temporal variance also means that processing time is different for computers and there are opportunities for digital agency where human agency is not possible due to the speed information travels. Building upon our autonomous vehicles example, some technology pioneers such as Elon Musk suggest that one day human-operated vehicles will be illegal, noting that "computers will do a much better job than us to the point where, statistically, humans would be a liability on roadways" (Lowensohn, 2015, para. 1). Since computers can process so much information and sensory data rapidly and in an objective fashion, they are useful for these types of emergency situations. However, the very idea of relinquishing control to a computer in life-or-death situations like driving an automobile is a scary prospect for many human operators, even if those actions end up saving lives most of the time.

A second related aspect of time made different by the digital is the *duration* and *permanence* of digital information. As we discussed in earlier chapters, one of the algorithmic and procedural characteristics of digital media is its ability to be copied exactly, in numerically perfect form, from one computer to another. This ability to "carbon copy" digital content has implications for all types of digital texts, from the documents we produce and consume on the Internet to complex software such as videogames and desktop applications. For example, consider the implications of bit-by-bit reproduction for online content. While many early works about hypertext focused on the seeming impermanence of the medium, there is another way to conceptualize hypertext and its associated digital objects, which is that these documents can live on for a very long time due to their ability to be easily copied, remixed, and shared throughout the Internet. Even if an original source is taken down or deleted, it is likely that another user has captured screenshots or images of the material or that the material has been harvested by a web archive tool, such as the Internet Archive Wayback Machine (www.archive. org), which allows one to revisit sites as they existed at earlier points in time. In Justine Sacco's predicament described above, one of Sacco's friends tried to delete the Tweet once it was clear the message had gained notoriety. The friend "frantically deleted her friend's tweet and her account—Sacco didn't want to look—but it was far too late. 'Sorry @JustineSacco,' wrote one Twitter user, 'your tweet lives on forever'" (Ronson, 2015, para. 13). Even though the "real" tweet had been deleted, the copies lived on indefinitely, regardless of the original author's desire for them to be removed.

A final way in which digital technologies have re-situated agency through temporal means is by *directing our attention and control toward their maintenance*. This too speaks to content developers' desires for content to be sticky and for user interactions to be sustained. Often, this redirection seems like another relationship we must maintain, but the relationship is now virtual-to-human rather than human-to-human. Today's mobile games played on Android and iPhone devices, for example, often require players to login once a day or at specified time intervals in order to level up their characters, collect in-game tokens or gemstones, or otherwise accumulate virtual currency in the game (Grassi, Barigazzi, & Cabri, 2017). When these actions are *not* performed, other humans who are on teams with the player may encourage the player to login or may even become frustrated enough to harass the player. Similarly, digital notifications delivered through virtual banners, email chimes, and notification badges constantly remind their human users that there are things to do. These tasks may include responding to emails from other humans or directing users' attention toward updates in the news or weather or asking them to attend to configuration and maintenance tasks required by other virtual applications. These days, using mobile applications can be a full-time job,

As everlasting tweets, needy virtual pets, and nagging and flagging software applications demonstrate, our technological devices are not limited to the same biological processes that determine and support one's human lifespan. Instead, computational limitations are determined by the physical durability of the components—materials like silicon, wiring, controllers, and peripherals—that house the data. However, a significant difference is that computers can be upgraded with new components, keeping the core memory intact. One day we may be able to upgrade our own memories in this fashion, by simply adding new parts when the old ones wear out. But until then, what methods do we have available to preserve our legacies?

Ghosts in the Machine: Self-Contained Electronic Memorials

The ability to plan for and create electronic monuments or memories for one-self—to be displayed upon one's death—is an illustrative activity to consider in relation to digital technology, agency, and ethics. It seems reasonable to state that one's physiological and biological ability to control and exert power ends when one is laid to rest. However, do power and control continue after one's death? Is it possible that one can be thought of as having agency even after one dies? Are there ethical rights that exist beyond one's living relatives and how they feel about things after one dies? Certainly trusts and other legal documents exert agency over how one's posthumous resources are divided and allocated, but are there other examples of agency that move beyond legal matters?

Wilkinson (2002) argues that the deceased do in fact have "posthumous interests" and notes that "there is often a symmetry between the interests of the living and dead people that can guide us in working out how to protect the interests of the dead" (p. 31). He describes, for example, scandals such as the Liverpool children's hospital in the U.K. where "at least 104,300 organs, fetuses, and body parts were stored in hospitals ... many had been taken without the knowledge of relatives, and so, obviously, without their consent" (pp. 31–32) and controversies surrounding research studying "the remains of the long dead where it is either certain or very likely that they belonged to indigenous groups whose successors oppose the research" (p. 32). In considering such cases, Wilkinson suggests that our notions of posthumous interests should look beyond just the implications for still-living relatives and should also consider notions like privacy, memory, and reputation. What ethical obligations do we have, for example, to preserve the interests of the dead even when they may contrast with the interests of the living?

With this background in mind, consider U.S. Patent 6414663 B1 (Manross Jr., 1998), which proposes a description and method for a self-contained computational tombstone. The tombstone possesses a central processing unit (CPU), programmable memory, a power source (batteries or solar cells), and an electronic display. The patent provides multiple form factors for such an invention; in one model, the display is built into the tombstone itself, allowing the display of "text and digital photos/images on an LCD display relating to the deceased's life, accomplishments, philosophy, genealogy, favorite photographs, or whatever they would like that could be rendered digitally" (Manross Jr., 1998, para. 4). In another form, the device can be buried with the deceased inside a coffin, "so that centuries from now if the remains were ever disturbed, people would know not only who this person was, but what their life was like" (Manross Jr., 1998, para. 6).

By the year 2007, similar types of electronic tombstones were available for purchase, but they were not popular items according to many funeral vendors (Imrie, 2007). However, even these vendors acknowledged that some elderly clients acknowledged the potential of such devices. As one sales vendor explained, "I see no reason … why I couldn't stand in front of a video camera and give a message to my grandchildren, such as: 'Faith in the Lord was important to me'" (Imrie, 2007, para. 27). Such digital monuments offer new functionality to preserving memories and memorializing the deceased—one could play a deceased relative's favorite musical passage, display favorite photos or audio files of the deceased, or, theoretically, even answer questions programmed into the system's online database while the person was still living. External sensors can extend the capabilities of such systems to also consider the environment in which the tombstone is embedded. For example, a cemetery in Slovenia created a digital tombstone that according to a news headline "blends multimedia with the macabre" by using a motion sensor mounted to a traditional style headstone to activate a multimedia screen when the tombstone is approached by visitors (NBC News, 2017). It is also easy to imagine a system that uses biometric sensors or password protection so that information is only accessible to living relatives with the proper access credentials.

Given Wilkinson's ideas about posthumous interests, the example of self-contained electronic memorials above, and what we have discussed in relation to agency and digital technology in this chapter, consider this hypothetical situation: You have been asked by your mother, who is not what you might call a "technical person," to prepare an electronic monument to your father, who has recently passed away. Your mother has asked for a slideshow of iconic photographs from your father's past. After reading an article in *Time Magazine* about this subject, she now also wants to embed a QR code so that visitors to the tombstone can use their cameras to pull up a memorial web page and add comments and memories to this online space. However, you know your father was an intensely personal and private man who valued his reputation in the community. You worry that such a monument may trivialize the sanctity of his memorial gravestone and you are not happy with the lack of control you would have over the QR-code web forums. Who knows what might be posted here? Your father had never mentioned wanting something like this and you worry your mother is being too aggressive in asking for something you don't believe your father would wish to have, if he were still alive. However, your mother insists that your father was such a modest man that you would never truly know his impact on the world without giving the people who visit his memorial the opportunities to tell their stories about their interactions with him. And she needs your help to do this, since you understand the technology much better than she does.

In this scenario, how might the electronic monument violate the posthumous interests of your father? How is agency re-situated from your mother to you in this example, and what role does technology play in that re-situation? How are both temporal and social factors impacted by technology, and what do those shifted relationships mean for the agency of you, your mother, and your deceased father? In what ways are agency and autonomy shifted or extended? More generally, what questions should be asked, and which safeguards might be put in place, to ensure that a deceased individual's electronic legacy does not violate how that person felt about privacy when they were alive? Are there other ethical complexities that emerge from this thought experiment that shed light on the questions about power and control raised in this chapter? Such questions are useful ones to ask when considering the relationships between digital technologies and human agency and autonomy.

Next Up

We next move our discussion toward the practical implications of the various topics we have been discussing throughout the book. We will present various frameworks for motivating action in the next chapter. This considers action in different contexts—political, administrative, and community—and provides guidelines from professional organizations that may be useful in considering digital ethics within specific professional contexts.

References

Apple, Inc. (2017). App store review guidelines. Retrieved from https://developer.apple.com/app-store/review/guidelines/.

Archive.org. (2017). Internet Archive Wayback Machine. Retrieved from http://archive.org/web/.

Bakshy, E., Messing, S., & Adamic, L. A. (2015). Exposure to ideologically diverse news and opinion on Facebook. *Science, 348*(6239), 1130–1132.

Boler, M. (Ed.). (2008). *Digital media and democracy: Tactics in hard times.* Cambridge, MA: MIT Press.

Brooks, D. (2017, November 20). How evil is tech? *The New York Times.* Retrieved from https://www.nytimes.com/2017/11/20/opinion/how-evil-is-tech.html.

Buckingham, D. (Ed.). (2008). *Youth, identity, and digital media.* Cambridge, MA: MIT Press.

Cash, H., D Rae, C., H Steel, A., & Winkler, A. (2012). Internet addiction: A brief summary of research and practice. *Current Psychiatry Reviews, 8*(4), 292–298.

Chen, C., & Leung, L. (2016). Are you addicted to Candy Crush Saga? An exploratory study linking psychological factors to mobile social game addiction. *Telematics and Informatics, 33*(4), 1155–1166.

Dibbell, J. (1993, December 23). A rape in cyberspace–How an evil clown, a Haitian trickster spirit, two wizards, and a cast of dozens turned a database into a society, *The Village Voice.* Retrieved from https://www.villagevoice.com/2005/10/18/a-rape-in-cyberspace/.

Dworkin, G. (1988). *The theory and practice of autonomy.* Cambridge, UK: Cambridge University Press.

Eltantawy, N., & Wiest, J. B. (2011). Social media in the Egyptian revolution: Reconsidering resource mobilization theory. *International Journal of Communication, 5*, 18.

Emirbayer, M., & Mische, A. (1998). What is agency? *American Journal of Sociology, 103*(4), 962–1023.

Floridi, L. (2013). *The ethics of information.* Oxford, UK: Oxford University Press.

Grassi, D., Barigazzi, G., & Cabri, G. (2017, September 22). User longevity and engagement in mobile multiplayer sports management games. Retrieved from https://arxiv.org/pdf/1703.03831.pdf.

Imrie, R. (2007). So far, high-tech tombstone has few takers. *NBC News.* Retrieved from http://www.nbcnews.com/id/22207941/ns/technology_and_science-innovation/t/so-far-high-tech-tombstone-has-few-takers/.

Kamvar, S., Haveliwala, T., & Golub, G. (2004). Adaptive methods for the computation of PageRank. *Linear algebra and its applications, 386,* 51–65.

Kominers, S. D. (2009). Sticky content and the structure of the commercial web. In *2009 Workshop on the economics of networks, systems, and computation (NetEcon '09),* Stanford, CA.

Lawton, L. (2017). Taken by the Tamagotchi: How a toy changed the perspective on mobile technology. *The iJournal: Graduate Student Journal of the Faculty of Information, 2*(2), 1–8.

Lotan, G., Graeff, E., Ananny, M., Gaffney, D., & Pearce, I. (2011). The revolutions were tweeted: Information flows during the 2011 Tunisian and Egyptian revolutions. *International Journal of Communication, 5,* 31.

Lowensohn, J. (2015, March). Elon Musk: cars you can drive will eventually be outlawed. *The Verge.* Retrieved from https://www.theverge.com/transportation/2015/3/17/8232187/elon-musk-human-drivers-are-dangerous.

Manross Jr., D.N. (1998). Self-contained electronic memorial. *U.S. Patent 6414663 B1. Google Patents.* Retrieved from https://www.google.com/patents/US6414663.

Medeiros, J. (2015, January). How Intel gave Stephen Hawking a voice. *Wired.* Business. Retrieved from https://www.wired.com/2015/01/intel-gave-stephen-hawking-voice/.

Murphy, M. (2017, August). Apple's 'walled garden' approach to content has paid off massively. *Quartz.* Retrieved from https://qz.com/1045671/apples-walled-garden-approach-to-apps-and-music-has-paid-off-massively-aapl/.

NBC News (2017, April). This 'digital tombstone' blends multimedia with the macabre. *NBC News Video.* Retrieved from https://www.nbcnews.com/video/this-digital-tombstone-blends-multimedia-with-the-macabre-919562819843.

Ronson, J. (2015, February 15). How one stupid tweet blew up Justine Sacco's Life. *The New York Times.* Retrieved from http://www.nytimes.com/2015/02/15/magazine/how-one-stupid-tweet-ruined-justine-saccos-life.html.

Schlosser, M. (2015). Agency. *Stanford Encyclopedia of Philosophy.* Retrieved from https://plato.stanford.edu/entries/agency/.

Turkle, S. (2011). *Alone together: Why we expect more from technology and less from each other.* New York, NY: Basic Books.

Vincent, J. (2016, March 24). Twitter taught Microsoft's AI chatbot to be a racist asshole in less than a day. *The Verge.* Retrieved from https://www.theverge.com/2016/3/24/11297050/tay-microsoft-chatbot-racist.

Wilkinson, T. M. (2002). Last rights: the ethics of research on the dead. *Journal of Applied Philosophy, 19*(1), 31–41.

9

DIGITAL AND ETHICAL ACTIVISM

Becoming more informed about digital ethics, and about both the obvious and subtle shifts in our thinking and behavior when we engage with technologies, is only the first step in understanding the moral complexities of our digital world. If an individual wants to do more with this information, perhaps aspiring to contribute to a more equitable society, then she or he must apply that knowledge in a productive fashion, informing decision makers, helping draft better policies, or catalyzing change through community action. We insist throughout the book that for a digitally and morally literate person, action must follow as the result of sensitivity, reasoning, and motivation. However, this action does not need to occur in a vacuum; there are resources that can be drawn upon for assistance. Example of these resources include policies that have been developed by professional organizations for guidance and best practice and strategies for digital activism from which we can learn. Consider this example of local community activism where impact and reach were greatly expanded by digital technologies:

CASE STUDY: THE DAKOTA ACCESS PIPELINE AND SOCIAL MEDIA ACTIVISM

In 2014, the $3.8 billion Dakota Access Pipeline (DAPL) project was announced to the public: an underground oil pipeline routed through four states (North Dakota, South Dakota, Iowa, and Illinois). The DAPL covered nearly 1,200 miles of territory (Hersher, 2017) and was expected to carry 570,000 barrels worth of crude oil daily (McQueen, 2018). However, the project was soon embroiled in controversy, with significant community opposition during its construction in 2016 and ongoing legal disputes and protests through early 2017. Specifically, one crossing of the pipeline, a segment across the Missouri River (Lake Oahe), was contested by the Standing Rock Sioux who argued the route "would jeopardize the primary water source for the reservation, and construction would further damage sacred sites near the lake, violating tribal treaty rights" (Hersher, 2017, para. 3).

What began as a series of relatively small-scale protests eventually developed into "the biggest coordinated Native American protest in history" (McQueen, 2018, p. 51). Using social media and its ability to draw global audiences into news and discussions surrounding the DAPL, the protest grew to include many other participants beyond the original Native American citizens, including many non-native protesters. The #NoDAPL hashtag united protestors on Twitter and activists leveraged their mobile smartphones to produce videos and distribute them on YouTube. Ultimately, the protests evolved into a movement referenced by the #NoDAPL hashtag.

The #NoDAPL case highlights how communities can use digital social media both to band together to confront issues of importance—in this case, a social injustice against indigenous tribes and the larger environmental problems associated with the DAPL—as well as be threatened by cultural appropriation facilitated by those same digital technologies. The case also shows how viral distribution can give an immediate and powerful platform for protesters to bring attention to problematic discourse or behavior, as occurred with media attention surrounding the use of water cannons on protesters in below-freezing temperatures at a rally in November 2016. In this case, social media was used to contradict a statement released by police that the cannons were being used to put out fires rather than as a crowd control mechanism (Johnson, 2017).

The DAPL project was ultimately seen by protesters as one that was pushed through the production and development process without proper communication and feedback by the local communities affected by the pipeline.

Challenges of Motivating Action through Policy

When bureaucratic and legislative procedures are found insufficient by the people controlled by them, digital technologies are increasingly being used to apply pressure to the decision makers involved in these projects. Policy is one avenue available for us to build possibilities for change, but as the DAPL case reveals, policy does not always satisfy the people governed by it, nor are policymakers always inclusive enough of their audiences. Two additional challenges relate to enforcement and equity. Since policies outline appropriate behaviors and specify the acceptable conditions for activities, they can be useful tools to regulate actions within a specific set of norms. However, these policies need to be enforced to be effective. For instance, if an organization developed a policy to enforce fair working conditions, free of harassment and discrimination, and then failed to follow through with punitive actions for those who violated that policy—or failed to develop preventative measures to discourage them from happening in the first place—then these guidelines would not be useful for actually creating those fair working conditions. Similarly, if the policy is actively enforced, but only for certain groups of individuals within the organization, such as non-executive employees, then the policy has done something even worse. It has introduced double standards to ensure that only those individuals with insufficient financial resources or those lacking executive authority will be held to this standard.

Given our current digital landscape, many of these policy issues are further exacerbated. For example, another challenge—which we have discussed throughout this book—relates to the speed at which information, greased by digital technologies, is produced and disseminated throughout society. It is commonplace for policies involving the digital to fall behind the practices and behaviors they are designed to eventually regulate. Because computing and network technologies sprint forward at a speed that exceeds bureaucracy's ability to match pace, we are often left with fundamental gaps in our ethical understanding of how digital data and functionality impact our lives, how our actions within digital networks impact others, and therefore how best to regulate these developing digital technologies. Sometimes digital processes become embedded in our everyday leisure activities, as mobile smartphones and social media technologies have done over the last decade, in ways that change our behaviors, actions, and contexts.

One example of this is bringing computing devices into places they were formerly not used, like a public park. Such seemingly subtle shifts in access and usage introduce new technological, social, cultural, or economic challenges into those spaces. For example, resource management, privacy, surveillance, and safety are all issues that are impacted when using mobile computing devices and remote-controlled drones in public places. These issues can range from dealing with unexpected traffic loads on public Wi-Fi access points to protecting the privacy of families who do not wish to be recorded by mobile smartphones or drones equipped with high-definition cameras. Without appropriate policies in place to dictate how these tools can be used, it is often a case of "anything goes" until something bad happens that requires a policy to be drafted. On the other hand, citizens are also empowered to become contributors to news and capture unjust activities on camera to help bring about change.

Similar mismatches between digital pace and policies occur in the corporate world. Consider health care providers' increasingly common use of electronic health care records. Goodman (2017) suggests that our ethical understanding of healthcare information technology needs to be improved and used to develop new policies around their construction and use. While health care practitioners and hospitals have been moving toward the digitization of health care records and medically captured information for years now, we have yet to fully understand the implications of these movements from analog to digital records. Such information describes and quantifies our bodies, our maladies, and our histories, and the implications of making such information digital are still being theorized. For example, philosophers such as Luciano Floridi stress the importance of developing a philosophy of information that acknowledges that computing technologies and online data play a role in our constructions of self (2013, 2017). In Floridi's view, our sense of self and identity draws not only from our biological interactions with our physical environment and our psychological impressions of these relationships, but also from our conceptualizations of identity in online and digital spaces (another topic we discussed at length in Chapter 6).

There are also chronological factors at work here that relate to digital agency as we have discussed previously in Chapter 8. Digitized healthcare information is an extension of us both as we exist in the present and as we have existed in the past. It also speaks to the potential trajectories and statistical possibilities of our state of health, wellness, or disease in the future. Practically speaking, these chronological variables are packaged digitally as a reflection of our biological and psychological selves that is easily accessible, sharable, and replicable. Such packaging may create ethical conundrums, if not sufficiently addressed with enforced policies. For example, we can imagine significant implications regarding the facility and ease with which records can be shared with insurance companies. This brings troublesome possibilities to mind regarding insurability for those with prior histories of illness or genetic markers which suggest a predisposition toward major (and expensive) illnesses in the future, such as cancer. However, there are also potentially advantageous possibilities with respect to ease of maintenance, backup, and cost savings. We can apply different ethical paradigms to these digital systems to recommend different types of policies or to yield different moral questions that must be considered by healthcare providers, patients, insurers, technology developers, caregivers, and the many other stakeholders involved in these digital relationships and transactions.

Despite the myriad challenges that accompany policymaking for digital environments, we can find numerous examples of policies that prove effective for regulating activities in digital spaces. Many professional organizations, for example, have developed policies for the ethical use of digital information and for the proper use of digital technologies within the context of their fields. In the next section, we will consider examples taken from different professional policies to analyze how different organizations have addressed digital challenges within their domains.

Policymaking in Research and Professional Organizations

Because they expect their employees to uphold high moral standards and represent the values of their disciplines at large, many professional organizations develop standards for ethical conduct and research that pertain to their scope of business connected with digital environments. In fact, the difference between a profession and an occupation is defined in the public administration literature as not just a degree of rigor and specialization in training, but also "by a commitment to a set of ethics and an obligation to serve faithfully" (Goss, 1996, p. 575). In addition to understanding the core knowledge and practical dimensions of a field, professionals are also expected to understand the cultural context of the field and be aware of their moral obligations when interacting with their stakeholders. For example, a physician with deep knowledge about anatomy and physiology but who was ignorant of the Hippocratic Oath (modernized in the AMA Code of Medical Ethics) would not be accepted as a morally competent medical practitioner in the profession. In addition to knowledge and skill, medical professionals are also expected to be familiar with ethical practices for interacting with patients, for treating those in need, and for ensuring that any treatment provided does more good than harm.

Stakeholder relationships in professional contexts are complicated by digital tools and researchers seek to better understand how and where these relationships are being changed. Researchers working to study the Internet have begun to address some of these issues in their own ethical policymaking. For example, the report produced by the Association of Internet Researchers' (AOIR) Ethics Working Committee is perhaps the most comprehensive digital policy document we have available on ethics and internet research (Association of Internet Researchers, 2012). This document notes specific tensions introduced by digital environments in considering the division between public and private spaces as well as our relationships to our digital identities (and whether or not the collection of data taken from our activities in online communities causes potential harm to our real world lives). In their policy document, the authors note,

> The uniqueness and almost endless range of specific situations defy attempts to universalize experience or define in advance what might constitute harmful research practice. We take the position that internet research involves a number of dialectical tensions that are best addressed and resolved at the stages they arise in the course of a research study. In saying so, we reiterate the value of a casuistic or case-based approach.
> *(Association of Internet Researchers, 2012, p. 7)*

Such a case-based approach acknowledges the numerous varieties of site-specific issues that may emerge and also highlights the fact that there may be tensions between stakeholders that complicate these scenarios. For example, the usefulness of findings may be diminished if too much personal information is removed from a data set; in this case, an ethical obligation to protect user privacy may be in direct conflict with an ethical obligation to disseminate useful research to the broader research community. It is a utilitarian dilemma, to be sure.

In addition to ethical codes for research conducted in digital environments, there are also ethical policies in place to guide medical and psychological professionals in their professions. For example, although the American Medical Association's Code of Medical Ethics (AMA, 2016) does not explicitly mention "digital" or "electronic" anywhere in its document, it does specify principles that implicate ethical digital tool usage in a number of ways. For example, Guideline V in the code calls for a physician to "study, apply, and advance scientific knowledge, maintain a commitment to medical education, make relevant information available to patients, colleagues, and the public, obtain consultation, and use the talents of other health professionals when indicated" (AMA, 2016, para. 6). One can easily imagine how many of these activities are supported through tools like online educational repositories, electronic data storage systems, email, FaceTime consultations with other physicians, and so forth. One can also imagine the need for precautions and safeguards around electronic data with confidential and protected patient information, as is outlined in the Health Insurance Portability and Accountability Act (HIPAA) guidelines, which are available for review on the U.S. Department of Health & Human Services website (HHS, 2017). Clearly, ethical conduct in the profession depends on ethical conduct with digital technologies. Similarly, the Ethical Principles of Psychologists and Code of Conduct (APA, 2017) contains ten detailed sections outlining expectations for the ethical conduct of psychologists. Several of these, including Section 4 (Privacy and Confidentiality) and Section 6 (Record Keeping and Fees) have significant implications for professionals who use digital technologies to facilitate their patient visits.

There are also ethical standards derived for technology design and development disciplines, such as engineering. For example, the National Society of Professional Engineers (NSPE, 2017) developed a Code of Ethics that outlines the ethical obligations for engineers as they fulfill their professional duties. Their policy notes that contracted designs cannot be duplicated by engineers without written permission and that the safety, health, and welfare of the general public should be held paramount. These ethical guidelines are complicated by both the properties of digital media, such as the ease and efficiency in which digital information can be copied and distributed, and the applications of digital media involving public safety, such as the autonomous vehicles and robotic health companions discussed in previous chapters.

Policymaking for Individuals and Communities

While it is useful and appropriate for professional organizations to develop ethical policies and codes of conduct that they expect members in their community to follow, the ethical landscape is further complicated by many ordinary citizens who have computers in their pockets and the ability to interact in environments formerly limited to professionals from those fields. Journalism is a great example of this, and we discussed in Chapter 8 everyday citizens using their newfound digital agency to become bloggers and cyberactivists. However, everyday citizens may be ignorant about these ethical policies given that they were not trained in these disciplines and do not attend the same conferences and professional meetings in which the policy would be discussed, created, and voted upon. In these situations, new community-moderated policies that take input from both everyday citizens and professionally trained journalists may be more helpful.

As Stephen Ward at the University of Wisconsin–Madison's Center for Journalism Ethics writes, "this new mixed news media requires a new mixed media ethics – guidelines that apply to amateur and professional whether they blog, Tweet, broadcast or write for newspapers. Media ethics need to be rethought and reinvented for the media of today, not of yesteryear" (Ward, n.d., para. 6). In other words, because of the newly complicated spaces between public and professional discourse, we cannot rely purely on ethical standards from professional organizations, even if we were to assume they were uniformly followed and respected. However, we can draw from such policies as a starting point to think about these ethical issues as they have been articulated by professional groups and organizations. It is also true that we can look specifically at how they have chosen to deal with new challenges introduced by digital technologies.

Policies for the Digital

In addition to ethical guidelines written specifically for conduct and research within digital spaces, like the Internet, there are also policy guidelines that have been drafted for discipline-specific uses of digital technology. When considering new policies for a community group or professional organization, it is worth perusing standards from these existing organizations rather than reinventing the wheel. For one thing, it saves a great deal of cognitive and physical work since the policies are often generalizable, even if they may require some minor adjustments. For another, it is useful to learn from the historical incidents that caused these policies to be developed in the first place. Sometimes policies are drafted due to something that has gone wrong in the past and it is often more palatable to learn from others' mistakes than from one's own mistakes.

Consider the following excerpts from the ethical codes of conduct of different professional organizations. Each of these policies deals with particular aspects and properties of digital media. Although many of these policies overlap with one another in both form and content, we have extracted just a few sentences from each in order to highlight how these policies have chosen to deal with some of the properties of digital media that we discussed in Part Two of this book.

- Regarding the authenticity, reliability, and privacy of electronic data, the American Psychological Association (APA) Ethical Principles of Psychologists and Code of Conduct (APA, 2017) calls for accuracy and the avoidance of deceptive statements in both print and electronic materials. It also mandates that clients who work with psychologists using electronic means are advised to "the risks of privacy and limits of confidentiality" (p. 7) in this medium.

- Regarding parameters for searching electronic data, verifying identity, and ensuring equity of access, the National Association of Social Workers (NASW) Code of Ethics (NASW, 2017) outlines a number of specific provisions for working with electronic data. Among other things, social workers are advised of their ethical obligations to not perform electronic searches on clients without their consent, to verify clients' identities and locations when communicating with them electronically, and to be aware of potential socioeconomic disparities when interacting with clients using electronic technologies.

- To combat potential skewing or manipulations of data digitally in reporting and photojournalism, The National Press Photographers Association (NPPA) Code of Ethics asks visual journalists to treat subjects with dignity and to "maintain the integrity of the photographic images' content and context" (NPAA, 2017, para. 6). The Associated Press features a detailed page on fabrications, nothing that they "don't stage or re-enact events for the camera of microphone, and … don't use sound effects or substitute video or audio from one event to another" (AP, 2017, para. 2).

- To expand our notions of research and data collection in online and digital environments, the Association of Internet Researchers' ethical guidelines clarify that we must think about the human subjects involved in electronic research even when "it is not immediately apparent how and where persons are involved in the research data" (AOIR, 2012, p. 4). For example, we might think of an ethnographic study involving how families use central space for communication but not consider that the digital photographs used as data might also include metadata with exact GPS coordinates to these families' homes.

- To protect consumer privacy and suggest safeguards for Internet users, the Digital Analytics Association's Web Analyst's Code of Ethics (DAA, 2017) outlines specific guidelines organized around five guiding principles: privacy, transparency, consumer control, education, and accountability. Education in particular is important to note in this policy, as it suggests a moral obligation for data analysts, who understand technology and Internet data, and asks them to educate staff and management about data collection practices, risk, and capabilities.

- To protect student and faculty privacy and provide accountability to taxpayers, many public universities have detailed policies outlining information technology usage on their campuses. For example, our home university has an Information Technology Policy that, among other things, outlaws the viewing of pornographic materials on university-owned computers and specifies that only the owner of each machine may access his or her email or voicemail except in very specific circumstances, such as criminal investigations, when others may also be granted access (University of Central Florida, 2017).

There are many additional examples of policies that provide language designed to shape behaviors in digital spaces. However, the range and diversity of just these few examples shows how prevalent digital technology has become in all of these fields. It is also interesting to see the overlap between these different policies since many digital considerations, such as user privacy, are critical ethical issues in many different professional contexts. All professional codes of ethics are adaptable—able to be updated by members of the communities they govern. Yet their adaptability is contingent on the ethical literacy of their professional members, so are often slow to reflect changes in technologies and digital habits. Actions by ethically literate professionals can stimulate the ongoing maintenance of codes of ethics and provide ethical leadership to their communities.

In the second half of this chapter, we consider some of the ways in which our digital technologies motivate action, either explicitly or unintentionally. These include politics, administrative work, and personal practices. For those wishing to advocate for particular causes or expose wrongdoings that our digital devices can more easily capture, we discuss policies that have been developed for the purpose of collecting digital evidence that can be used in courtrooms and legal proceedings. We conclude with a case study of a volunteer organization that used digital techniques to advocate for an environmental campaign to remove a gravel pit from a local community.

Motivating Action through Practice

While policy documents are useful for outlining expectations for appropriate behaviors, there are also opportunities to effect change through more active means. Traditionally, crusaders for various causes have used measures like protesting, writing letters to elected officials, penning opinion pieces for their local newspapers, or engaging local neighbors in conversation or debate to drum up support for their activities. With new digital technologies, these practices have expanded in both scope and scale. For example, in Chapter 8 we briefly discussed the work of cyberactivisits who use blogs, social media, and other electronic means to advocate for their causes. Here we explained how digital agency can sometimes provide a voice for the "everyday citizens" who lack the power to introduce bills into legislation or mobilize large political parties. Since cyberactivism draws upon the properties of digital media we have discussed throughout this book, it can be a powerful means of motivating action and distributing time-sensitive information quickly and efficiently to audiences worldwide.

In addition to these more purposeful activities, there are also moments when action is catalyzed by digital documentation without any detailed planning or premeditation. For example, Ess (2014) explains that our access to digital technology moves us out of the realm of consumer and into the mode of producer; this opens up moments of serendipitous discovery and unplanned possibility for documenting the world around us. For example, one may purposefully seek to catalyze action by penning an online column or blog posting about a current event with the specific aim of changing opinions about or building support for a particular politician or cause. On the other hand, the ubiquity of everyday citizens with easy access to smart mobile phones and recording devices also means that one may *unintentionally* play a role in motivating action by being in the position to capture dramatic footage as it is happening. We have seen examples of this with mobile footage of manmade or natural disasters, for example, or citizen footage of controversial shootings or human rights abuses. In these instances the persons capturing the footage just happened to have a recording device and be in the "right place at the right time" or perhaps the "wrong place at the wrong time," depending on your perspective.

As a result of both planned and unplanned recordings made by citizen journalists, new types of policy documents have emerged that specify how to capture footage that can be successfully used in courtrooms as evidence. For example, the "Video as Evidence Field Guide" produced by the Witness Organization is a detailed online guide that covers how video is used in and outside of the courtroom. It includes guidelines about how to "ethically and effectively use video to document abuses and support the process of bringing perpetrators to justice and freeing the wrongly accused" (Witness.org, 2017). This is an example of how a policy document was created in response to actions that were already occurring. The document was written and annotated with comic-style graphics and images in order to (a) make the videos that were being captured more useful as tools for bringing about change in the courtroom, and (b) make the policy guide as accessible as possible to a wide range of everyday citizens across the world. In addition to the accessible style of presentation, the field guide is also translated into multiple languages, including Arabic, English, Russian, French, Spanish, and Portuguese. This type of presentation is a good reminder for us as ethical policymakers to consider not only the content itself within the policy document, but also how it is presented and made accessible.

In addition to linguistic features such as translation, widely used policy documents presented online should also include features for differently abled users, such as captions for video and other accessibility guidelines outlined by best practices and regulations. One important policy document for digital accessibility in the U.S., for example, is the Section 508 standards (Section508.gov, n.d.). These government standards provide guidelines for both hardware and software that regulate telecommunications equipment and websites to provide better access for users with a wide range of capabilities. Such considerations for user experience can help move ideas from words on a paper to actions in a community, as guides such as the Witness.org field guide are intended to do.

Although we acknowledge the possibility of motivating unplanned action through digital media, the remainder of this section will explore actions that are intentional. There are many different areas in which action is motivated through digital means, but for the sake of brevity, we will explore three different categories: political, social, and administrative actions.

Political Action

The U.S. in recent years has been in a period of tense and divisive political angst, with opinions toward issues often sharply divided among party lines. Political partisanship is reflecting deep populist divides among our citizens. Some policy changes championed by certain politicians— like whether to repeal the Deferred Action for Childhood Arrivals (DACA) program, or how to deal with internal intellectual property theft, or how best to protect and use the natural resources that support technological development—can further polarize these communities. Other issues more directly relevant to the digital are not only similarly divided along partisan lines but also hot topics for demographics who are heavy users of Internet technologies, such as the complicated legislation involving oversight of telecommunications regulations for the Internet in 2015 (Edwards, 2015) and then again in late 2017 (Downes, 2017). Indeed, research shows that the "partisan divide" in the U.S. between critical policy issues dealing with immigration, national security, and environmental protection reached record levels during President Obama's tenure and the divide has since widened further under President Trump (Pew Research Center, 2017).

Regardless of the political drama that might emerge around the family dinner table and make it difficult to talk about issues in a reasonable way, it is true that one of the most significant ways in which one can move policy into practice is by becoming involved in politics. Leadership positions in local, state, or federal government provide opportunities for not only writing and passing bills in legislature, but also for building support for issues within communities. Successfully making a run for office is certainly a topic for another book altogether, but it is worth mentioning here that issues surrounding digital technology and the Internet, such as the telecommunications regulation example discussed above, are increasingly becoming important issues in political discussions. Becoming involved in politics, whether seeking a position as an elected official or starting out in politics more gradually, as a member of a candidate or office holder's campaign or office staff, is one significant way to help motivate action. Additionally, such activity can empower individuals to more broadly improve our society's understanding of digital media and potentially to draft and enact plans to use this technology for morally beneficial purposes rather than just for business and leisure. And there are many other less intimidating options for exploring political action, too. Two examples are serving as a legislative aide for an elected official or volunteering to work on the campaign of a politician you endorse.

Administrative Action

If seeking public office or working for elected officials leaves a bad taste in your mouth, another option is to work in an administrative position as a public servant or for a nonprofit organization. Although such roles are naturally connected to politics, they can serve as another route to more directly influence ethically sound practices or to champion certain causes in your community. Svara (2014) notes the similarity of nonprofit organization administration to government administration—both types of management deal with clients or the general public and require one to distribute resources to a clientele that is larger than the resources can adequately service. As a result, ethical practice is critical in the fair and equitable distribution of goods and services to the populations who need them.

Because public administration is essentially the implementation of organizational policy, it is an opportunity for someone to play a leadership role in the transition of policy documents from ideas to a set of practices. There are numerous ethical problems which can be addressed in civil service, both at a broad societal level and in government itself. Svara (2014) describes four layers of stakeholders typically at work within both government and nonprofit settings: organizations, administrators, political superiors, and citizens or clients. Sometimes, these relationships are complicated. For example, a nonprofit is likely to have a Board of Directors who play a role in providing advice and guidance about the future directions of the organization. Administrators must answer to the Board, but also to political superiors who may have their own constituents they must answer to at the next election. And all of these stakeholders may depend on donations from the public in order to keep the organization running, further complicating the priorities of the agency and the optics of how it is perceived by the public. An individual who can successfully thrive in such a complicated environment filled with competing priorities can do much good in helping to translate policy to action that serves a group or community.

Personal and Community Action

While working in politics or nonprofit administration will appeal to some ethical crusaders, it may be less exciting for others due to the prerequisite investments in time and energy required for those types of positions. It may be that politics moves too slow or that nonprofit leadership is too constrained by its external Board of Advisors. Or, perhaps the cause you wish to advocate for is localized so that support for it will not exceed the boundaries of your community. For individuals who are not interested in government or administrative service, there are still plenty of opportunities to motivate action in their communities. In these cases, it is possible to rely upon smaller, more flexible avenues of building support and catalyzing action for particular causes in the community. Further, it may be that these types of more focused and localized activities at the personal or community level may be the most efficient and rapid way to motivate change to address the issues of concern. Whether through small changes in personal behaviors that can serve as examples to others (like taking some time each day to "go dark" without digital connection, an exercise we ask some of our students to take up in classes), or connecting your local engagement to small community groups, personal action can be a direct and powerful tool to build ethical habits in the digital.

If one chooses to catalyze actions through these personal routes, it is important to be aware of the dynamics and challenges within personal/community hybrid initiatives. Community or self-organized activities often have fewer resources to draw upon and are self-financed individually or with donations from friends and families. In addition, generational challenges can emerge as older and younger adults may have different attitudes toward digital media and different levels of digital literacy. In their research analyzing the social media use of advocacy organizations, Obar, Zube, and Lampe (2012) note digital literacy gaps between younger, more digitally literate employees and the senior staff who are often making policy decisions. This digital literacy divide continues to pose broader challenges for the strategic use of social media technologies for social justice campaigns. Since community-organized projects often evolve organically and exist moment by moment, with little opportunity for formal planning, there should be provisions in place for dealing with conflict within the group. Additionally, even for small projects, a good marketing campaign is crucial so that the information is distributed in a way that supports the goals of the campaign. For every successful online video that "goes viral" in support of a social justice issue, there are numerous others that are viewed by only a small number of community volunteers and organizers.

Water Warriors: The Concerned Citizens of Brant

We can find many cases of individuals organizing to address a challenge in their community. Often, these individuals are catalyzed by something that impacts their lives directly and personally. Consider the Concerned Citizens of Brant (CCOB), a group of community activists living in Paris, Ontario. In early 2015, participants in this organization were concerned with the opening and installation of a new gravel pit over a groundwater reservoir in their community. Their self-described mission was "protecting our source water and environment now and for future generations" (Concerned Citizens of Brant, 2015). Community members interested in supporting the cause could subscribe to a mailing list to learn more or join the discussion online using the Facebook or Twitter accounts linked from their website. Ethically, they were concerned about the long-term consequences of the gravel pit to the environment and to the water supply their families were using for potable water.

While the early activities of the CCOB made good use of their network to organize a group of concerned individuals around a central concern impacting their community, they felt as though individuals in power were not paying them enough attention and their efforts were not making an impact. They recruited some community members to attempt to generate some excitement around their initiative, hopefully one that would make an impact with policymakers and government officials. Initially, a "Stop the Paris Pit" Twitter campaign seemed to be making good progress (Figure 9.1), generating excitement and positive momentum around the issue by recruiting new community members and likeminded environmental activists.

An interview with the designer of the Twitter campaign, Cassie McDaniel (personal communication, March 17, 2015), revealed some of the ethical issues surrounding its launch and reception by the CCOB membership. In an interview with McDaniel, she noted several challenging aspects of the social media campaign, particularly in regards to the stakeholders. First, she outlined the background of the initiative more clearly:

> *The Concerned Citizens of Brant (CCOB) have been trying to stop a gravel pit for a couple years, pursuing decision-makers like local Councillors and the Ministry of the Environment mostly through science and research and academic papers, rather than public pressure. Their backgrounds were in public health, so that makes sense. I spoke with a long-time community organizer (not a CCOB member) for some advice and he assured me that decision-makers didn't care about the science – they cared about what their constituents were making noise about. So I wanted to use social media to make some noise.*

This summary provides evidence of social media's power to "make noise" about issues that will catalyze public action and, ideally, create enough activity so that politicians pay attention to what those community audiences are saying. With CCOB, it was clear the scientific evidence showing the environmental problems

FIGURE 9.1 Stop the Paris Pit (Twitter)

associated with these gravel pits was not persuasive or "loud" enough to mobilize appropriate actions. Organizers then turned to social media to try and cultivate a more impactful communication strategy. After describing the project's background, McDaniel proceeded to further explain the context in which the project unfolded:

> *The problem with doing this was that the group is mostly elderly volunteers (as retirees, they are some of the few people who have time to get involved with issues like this), and very few of them understood the way social media works.*
>
> *I posted a video on Facebook of the group doing a "CCOB wave" around the boardroom table after an invigorating meeting one evening. About a week later, I got some fuming emails saying that they had received a complaint from a CCOB member. I'm STILL not sure the nature of the complaint, but I suspect they thought the video was unprofessional and undermined their 'scientific' stance. They seemed to miss that I was trying to reach a new crowd with that video, to show people that activism could be fun, that you could have personality, that you don't have to be a stodgy old person to be an activist!*
>
> *There were several mentions about the Gravel Pit company's "spies" and how content like that could be used against us. So then there was a move to have everything pre-approved by the board before posting it to social media. That was when I stepped away – that is WAY too cumbersome a process, and took away any ownership I had to do what I do.*
>
> *It's a tricky line to walk when you're trying to fight something like that, because it's generally all volunteers – you want to do what is for the good of the group and the cause, but if you alienate your volunteers or don't trust their expertise, you risk losing their involvement and support. I've seen this happen in other volunteer groups too. So, it can be frustrating on both ends.*

Here, we begin to understand more about the practical challenges associated with a campaign and call to action using social media. First, there was the primary stakeholder audience, an organization composed of elderly retirees, many of whom did not possess a sophisticated knowledge of technology and the cultural conventions of social media and the viral videos that can "make noise" and activate or entice a younger generation of supporters. In particular, the quirky strangeness that often makes such viral hits so difficult to describe and predict was not seen as a positive attribute of the message, but rather as a damaging blow to the organizational ethos of CCOB.

Second, and perhaps more importantly within the context of ethics, there was mention of the issues of resource scarcity and the difficult and somewhat tenuous relationship often found between community participants and organizational goals. As we see in ethical situations involving diverse groups of stakeholders, there is often variance in how different individuals relate to complex notions like justice. With CCOB, there was similar disconnect between one group's expectation of optimal outcomes versus another's. In this case, a scarcity of resources led to asymmetrical goals. There were no financial resources to hire a professional social media firm to make the noise that CCOB found necessary. Accordingly, members of the board relied on volunteers to use their own skills with Facebook and Twitter to try and engage outside audiences.

Ultimately, this community advocacy led to problems for both established (and generally older) volunteers and for the newer (and generally younger) volunteers who were digitally savvy. The CCOB board members felt embarrassed by a perceived lack of professionalism and threatened by a lack of control over their message's distribution, while the newer volunteers felt disenfranchised and disappointed that their efforts were not seen in the same positive light as they had imagined. Ultimately, this led to a fracturing of the community, with CCOB volunteers declining to continue participating within the parameters outlined by the Board.

One of the interesting outcomes of the CCOB community activism case is the evidence of a gap between demographics in terms of how to effectively and ethically use social media for activist purposes. Overall, while it would seem as though the CCOB campaign was not successful, one aspect of social media is that even sub-optimal outcomes can produce positive forward momentum. In this case, the media attention that the group so desperately craved was in fact beginning to take root even after the disastrous "CCOB wave" incident described in McDaniel's interview. She further explains:

> There were some good things that came out of getting the group on Twitter – we were followed by several activists in other townships who were fighting similar battles. There was a tweet from the Ministry of Environment, Glen Murray, about clean water and all these people piggy-backed on each other's messaging to make a kind of snowball effect. There was also a young politician, who through Twitter, was extremely supportive of our cause, and whose support led to other media opportunities. And other groups that had successfully deterred gravel pits in their communities were opening up to us and giving us all sorts of advice.

Tweets from Murray (2015) and Food and Water First (2015) confirm the success of the initiative in gaining wider attention outside of CCOB. However, a careful review of this scenario suggests missed opportunities on both sides of the campaign. On the one hand, a deeper investigation of the organizational values and professional ethos of the CCOB group through a formal stakeholder audience analysis as well as the implementation of social media guidelines would have likely revealed important and useful information about the group's preferred communication style and organizational goals. Such an investigation might have yielded useful insights about the appropriateness of the CCOB wave given the nature of the group and their perceived stature in the community. On the other, a better understanding of digital media by the CCOB membership would have provided the group with a better sense of the nature of social media, viral videos, and audience engagement using the Web.

CASE STUDY: STAKEHOLDERS AND ACTIVISM

Using CCOB or another community activism case you find online as an example, consider how demographics such as age, ethnicity, and socio-economic background might play a role in how different social media strategies are deployed. How might you identify the various stakeholders involved in such a promotion, and what unexpected audiences might be affected by your work? What ethical factors should be considered when developing a strategy for community activism? Would a specific communications policy help to counteract some of the difficulties experienced by the CCOB volunteers and similar groups of volunteers for other organizations? If so, what types of items would be important to include in such a policy? Finally, what other political, administrative, or personal avenues for action might one pursue in advocating for a social justice or environmental cause?

Next Up

Throughout this book, we have gradually moved from more theoretical explorations of digital ethics in beginning chapters to more applied analyses of how ethical issues emerge digital contexts. In our conclusion, we discuss the implications of living in a digital world and consider future areas in which we hope this type of research will continue.

References

American Medical Association. (2016). AMA code of medical ethics. Retrieved from https://www.ama-assn.org/sites/default/files/media-browser/principles-of-medical-ethics.pdf.

American Psychological Association. (2017). Ethical principles of psychologists and code of conduct. Retrieved from http://www.apa.org/ethics/code/ethics-code-2017.pdf.

Associated Press. (2017). Fabrications. Retrieved from https://www.ap.org/about/news-values-and-principles/telling-the-story/fabrications.

Association of Internet Researchers. (2012). Ethical decision-making and internet research: Recommendations from the AOIR ethics working committee (version 2.0), 1-19. Retrieved from https://aoir.org/reports/ethics2.pdf.

Concerned Citizens of Brant. (2015). Retrieved from http://ccob.ca.

DAA. (2017). The web analyst's code of ethics. Digital Analytics Association. Retrieved from https://www.digitalanalyticsassociation.org/codeofethics.

Downes, L. (2017, November 27). Dear Aunt Sadie, please step back from the net neutrality ledge. *Forbes*. Retrieved from https://www.forbes.com/sites/larrydownes/2017/11/27/dear-aunt-sadie-please-step-back-from-the-ledge-on-net-neutrality/#1770c7a67d6c.

Edwards, H.S. (2015, March 13). Why 2016 republicans oppose net neutrality. *Time*. Retrieved from http://time.com/3741085/net-neutrality-republicans-president/.

Ess, C. (2014). *Digital media ethics*. Cambridge, UK: Polity Press.

Floridi, L. (2013). *The ethics of information*. New York, NY: Oxford University Press.

Floridi, L. (2017). Why information matters. *The New Atlantis, 51*, 7–16.

Food and Water First. [FoodWaterFirst]. (2015, January 18). Congrats on fighting the good fight! Keep us in the loop & we'll share. [Tweet]. Retrieved from https://twitter .com/FoodWaterFirst/status/556917023423627266.

Goodman, K. W. (2017). Introduction: Symposium on ethical issues in data science and digital medicine. *Cambridge Quarterly of Healthcare Ethics, 26*(2), 326–327.

Goss, R. P. (1996). A distinct public administration ethics? *Journal of Public Administration Research and Theory, 6*(4), 573–597.

Hersher, R. (2017, February 22). Key moments in the Dakota Access Pipeline fight. *NPR*. Retrieved from https://www.npr.org/sections/thetwo-way/2017/02/22/5149 88040/key-moments-in-the-dakota-access-pipeline-fight.

HHS.Gov. (2017). Summary of the HIPAA privacy rule. *U.S. Department of Health & Human Services*. Retrieved from https://www.hhs.gov/hipaa/for-professionals/privac y/laws-regulations/index.html.

Johnson, H. (2017). # NoDAPL: Social media, empowerment, and civic participation at standing rock. *Library Trends, 66*(2), 155–175.

McQueen, D. (2018). Turning a deaf ear to the citizen's voice. Digital activism and corporate (Ir) responsibility in the North Dakota access pipeline protest. In G. Grigore, A. Stancu, & D. McQueen (Eds.), *Corporate responsibility and digital communities* (pp. 5178). Cham, Switzerland: Palgrave Macmillan.

Murray, G. [Glen4ONT]. (2015, January 24). Can you imagine if your water supply was toxic like Toledo's was & boiling water didn't help. Impacts on health & local economies is huge. [Tweet]. Retrieved from https://twitter.com/Glen4ONT/status /559022390869757952.

National Association of Social Workers. (2017). NASW Code of ethics. Retrieved from https://www.socialworkers.org/About/Ethics/Code-of-Ethics/Code-of-Ethics-En glish.

National Press Photographers Association. (2017). Code of ethics. Retrieved from https://nppa.org/code-ethics.

National Society of Professional Engineers. (2017). Code of ethics. Retrieved from https://www.nspe.org/resources/ethics/code-ethics.

Obar, J. A., Zube, P., & Lampe, C. (2012). Advocacy 2.0: An analysis of how advocacy groups in the United States perceive and use social media as tools for facilitating civic engagement and collective action. *Journal of Information Policy, 2*, 1–25.

Pew Research Center. (2017, October). The partisan divide on political values grows even wider. Retrieved from http://www.people-press.org/2017/10/05/the-partisan-divide-on-political-values-grows-even-wider/.

Section508.gov. (n.d.). Section 508 standards. Retrieved from https://www.section5 08.gov/summary-section508-standards.

Svara, J. H. (2014). *The ethics primer for public administrators in government and nonprofit organizations*. Burlington, MA: Jones & Bartlett Publishers.

University of Central Florida. (2017). Use of information technology and resources. Policy 4-002.2. Retrieved from http://policies.ucf.edu/documents/4-002.2UseO fInformationTechnologiesAndResources.pdf.

Ward, S.J.A. (n.d.). Digital media ethics. University of Wisconsin-Madison Center for Journalism Ethics. Retrieved from https://ethics.journalism.wisc.edu/resources/digit al-media-ethics/.

Witness.org. (2017). Video as evidence field guide. Retrieved from https://vae.witness. org/video-as-evidence-field-guide/.

CONCLUSION: DIGITAL AND MORAL LITERACIES

If we have made one point firmly throughout this book we hope it is this: Becoming literate in both the digital and moral senses is as difficult as it is essential. Determining thresholds for moral imagination and background knowledge to enable ethical sensitivity demands continued and sustained engagement with questions about digital technologies, digital media, digital information, and the objects of a digital nature (in particular if that essence of the digital challenges "analog" ways of understanding human and nonhuman animal individuals). Reasoning through the topics and problems to which you then become sensitive is additionally hard: The competing and overlapping normative accounts of ethical pluralism ask you to balance and specify within specific and sometimes highly technical contexts. And, finally, actually *acting* on those reasons requires overcoming psychological, social, and interpersonal barriers, not to mention wading through bureaucracy and red tape, some of which we discussed in the previous chapter. Yet the work of literacy is, we think, essential work. Without it, we will not have the skills or insights to guide the direction of the rapidly changing, digitizing world. Without it, we cannot understand digital ethics.

One response to this claim of difficult but necessary work is to turn to policy and regulation, practical guides to action and value that someone else has worked out on our behalf. However, one thing surely missing in professional policies and regulations is reasons for their existence: *justification*. A list of rules and procedures is not enough to account for novel change, complex conditions, and nuanced contexts. Consider, for instance, Institutional Review Board (IRB) training that has become integrated into (at least) every research institution supported by federal funds and many if not most professional organizations doing research around the world. Ethical theories are embedded in that training protocol, but basically it has been presented as and governed by rules: Ethics has become *compliance*. Over the past three decades or so, core bioethical principles were added to those trainings, making them more robust and offering better justification for the rules they seek to enforce. But it is still possible for people simply to think of ethics policies and regulations as "checklist" approaches: where following rules or "codes of conduct" supports the bare minimum of what is morally required of us. To us, what is important here is that referring to a policy might provide a person with a reason or motivation to do or not to do something but that is different from offering a well-reasoned justification for the acceptance or rejection of a course of action.

We think that distinction is important. While acting in a certain way, avoiding breaking rules and ensuring that they are followed may be covered by a policy or regulation, it is not the case that ethics, including digital ethics, can be captured by a set of rules, a code of conduct, or regulatory guidelines. It is too complicated and richly thick. Regulatory compliance is not enough. Because of this complexity, ethics requires the development of moral and digital literacy which enable an individual to understand, to reason, to decide, and to act. Moral and digital literacy allow us to move beyond rules, policies, and procedures and to evaluate such structures and the situations and conditions to which they relate, and to accept, reject, or amend them as circumstances and developments permit or require.

Living in (Digital) Communities

One of the recurring themes of this book has been our focus on community. We have argued that a virtue theory approach is important because it is focused on and central to the concept of community, consistent with a pluralism of ideas, perspectives, theories, and problems. We have argued that the digital demonstrates more of the diverse ways in which we are each not merely isolated individuals, but richly interconnected and interdependent informational entities, whose reasons, actions, and inclinations are bound up in and through the communities with whom we interact and on whom we depend. And, we have argued that ethical issues in the digital are themselves not isolated or isolable: They demand collective interdisciplinary responses, like the ones we have come together to offer in this book.

We would be remiss if we did not take the opportunity to frame our project here in its larger community of scholarship: *Understanding Digital Ethics* is part of a quickly growing conversation about the ethical implications of digital technologies and digital culture that is taking place across numerous disciplines. Since we began thinking about this book, for example, several other volumes related to digital ethics have arrived. To us, this is exciting in that it demonstrates the relevance and growing interest in questions of ethics and the digital. One of these, Davisson and Booths' 2016 *Controversies in Digital Ethics* is an edited collection of essays that argues for the participatory turn in communication and increased responsibility of both journalists and consumers through media literacy. Their focus on controversies emphasizes the participatory community-based account of "ongoing disputes and heated conversations" (p. 3) and points to problems rather than solving them. We see our book as a strong complement to this more discipline specific take on digital ethics, as our effort here has been to theorize "the digital" writ large, beyond the context of any specific disciplinary account.

Another book, Elliott and Spence's 2017 *Ethics for a Digital Era*, offers a similar discipline-specific perspective from and on digital journalism ethics. The authors argue centrally that questions of digital ethics are just "old wine in new bottles" (2017, p. 4) and that traditional concepts apply traditionally. For example, the concept of "moral agents" continues to denote only the set of "competent, rational adults" (p. 4). Our approach differs here in that we cast a wider ontological net, open to catching more diversity of possible views about metaethical issues and, therefore, wider applied consequences. But, like our book, Elliott and Spence work across ethical and epistemic theory, applied ethics, and case studies. Their approach to build "self-reflection as an effective and efficient way for becoming a reflective media practitioner" (p. 8) parallels, in a narrower fashion, our account of moral literacy. Other volumes with more specific focus, like on rhetoric and responsibility in online aggression (Reyman & Sparby, 2019), on safe and legal behavior online (Leavitt, 2018), and on media ethics (Christians, 2019) all continue to evidence the important scope and scale of digital ethical issues.

Projects like these come together in a community of digital ethics to display, argue, and compare perspectives on the nature and implications of digital ethics. As we have argued throughout this book, our pluralistic approach carefully analyzes not only ethical issues through conceptual analysis and applied case studies, but also the nature of the digital itself. Pluralism remains open to a diversity of views, with the idea that together in community we come to better understand, more clearly value, and more quickly adapt. As we see it, our approach has three distinctive elements that advance our shared understanding of digital ethics beyond disciplinary-specific approaches and regulatory models.

The first key characteristic is a balance between accessibility and depth. Some of the existing literature around digital ethics is either too conceptually dense or too simply focused on technological issues. We have aimed to be more straightforward and applicable not just to philosophers or technology experts, but to *all* of our interdisciplinary readers who are likely already immersed in digital information contexts in their everyday lives. By providing access to complex ideas and arguments in digital ethics, we enable readers to more fully engage in the topic in ways relevant to their own backgrounds and issues.

We sought to discuss the key concepts and theoretical implications through specific cases, letting you, our readers, access the personal, professional, and societal implications of various decisions that are made in morally complex situations. By following this approach, we hope we have created a text that is useful for both theoretical work examining the new ethical questions and moral philosophies enabled by digital media, but also for practical reasons as a text useful in many different classroom environments (e.g., digital media ethics, philosophy of science and technology, Internet ethics, ethics of social media, etc.).

Secondly, in keeping in line with our goal for accessibility, our book offers an interdisciplinary approach to considering digital ethics, incorporating both theories and practical examples from a number of fields including philosophy, computer science, digital media, journalism, and media studies. As authors, our own disciplinary and experiential backgrounds come together to shape our argument about the nature and implications of digital ethics. In contrast to the work of Ess (2014); Couldry, Madianou, and Pinchevski (2013); Elliott and Spence (2017); and Christians (2019), our book does not emerge from a journalistic paradigm, but rather from considering the uniqueness of the digital and what that means for the human condition. We also incorporated, rather than shied away from, the technical details of digital media and discussed how those properties relate to ethics. These approaches required an interdisciplinary method for considering ethics. To integrate these ideas in a meaningful way, we articulated and defended a theory of digital ethics based on the core characteristics outlined above, exploring those in relation to case studies described in in each chapter.

Lastly, and perhaps most importantly, our book offers a timely update by drawing upon ethical cases from contemporary events in media and society, while at the same time offering useful tools for doing this same work on cases and problems that emerge after our book is in your hands. The rapid growth, adoption, and dissemination of digital information has produced and will continue to produce a great number of intriguing cases to consider from different moral viewpoints.

Case Studies and Argumentation

Throughout the book, case studies have been highlighted across chapter themes, but none of them have been "solved." That is, we take the approach that offering to readers some essential elements of reasoning and argumentation, including the importance of background research to provide the most accurate information for the best arguments possible, will allow readers to use the cases to apply ethical theories and principles and the elements of good reasoning to offer potential solutions to cases. This approach is necessarily bound to the conception of community of interest and concern that requires perspective-taking in the analysis of cases, positions and problems. It encourages people to work together to approach, analyze, and solve problems in digital ethics. The approach to analysis and argument, knowledge of theories and principles of ethics, and knowledge of the properties and characteristics of the digital—and applying them to cases— constitute moral literacy in digital ethics.

We feel passionately that expertise in ethics matters, so we have included an appendix to this book which lays out in specific detail a method for developing and analyzing case studies in digital ethics. The use of mind-mapping techniques, the division of labor in researching a case, and the careful application of sound principles of good reasoning we offer there can guide you beyond mere opinions and feelings to careful reasoning and conclusions.

The strong suggestion we provide in this book that, whenever possible, people work together in the attempt to solve problems or to offer *solutions* (i.e., not necessarily singular ones) mirrors the argument that a pluralistic approach to ethics (both with respect to the use of theories and with respect to expertise and points of view) is a valuable and effective means of attempting to solve problems. Following the lead of classical pragmatists in rejecting a search for and insistence on establishing Truth and accepting that we live in a world of truths (with a lowercase "t") allows us to conceive of possibilities that play themselves out in the "real world" and to recognize that there are often various means of understanding and of solving "real-life" problems. Such problems are in the practical realm, and, as John Dewey would have it:

> The realm of the practical is the region of change, and change is always contingent; it has in it an element of chance that cannot be eliminated. If a thing changes, its alteration is convincing evidence of its lack of true or complete Being. What is, in the full and pregnant sense of the world, is always, eternally. It is self-contradictory for that which is to alter … [and] [t]hat which becomes merely comes to be, never truly is…. The world of generation is the world of decay and destruction. Wherever one thing comes into being something else passes out of being.
>
> *(Dewey, 1929, p. 22)*

Dewey's sentiment here surely applies to digital technology given the rapidity and, as John Milton (1674) poetically put it in *Paradise Lost*, the "speed almost spiritual" by which the digital both changes and affects its users. Where the Commodore 64 was once the ultimate in home computing, it is now no more than a dinosaur. And our new Apple Watches and computers, Android devices, and Windows computers and operating systems (and the uses to which we put such things, and the problems we will encounter in their uses) will become obsolete, only to be replaced by more powerful and more complex machines and objects that we will use in our daily lives.

Also following the lead of more recent pragmatist approaches to problem-solving is to take Richard Rorty's view that we need to recognize our obligation to sympathize with the concerns and suffering of others and that we are able to discuss experiences, positions, ideas, and values to determine what we have in common. We can move on from there to find—as Dewey put it—solutions that "will do." In other words, we ought not to seek after certainty (see Dewey, 1929), but we ought to try to create a world that is better than the one in which we now live (Rorty, 1979). While we certainly do not think or write under any unreasonable, fanciful and decidedly non-pragmatic assumption that our book will make the world a better place all on its own, we do believe that the approaches we have taken in this book are a contribution to trying to do so in the world of digital technology and digital ethics.

Future Work

There are many different directions in which the work from this book might be extended. In some areas, such as the example scenarios and thought experiments we embedded into most of our chapters, we merely brushed against the surface of these cases to inspire deeper thinking about some of the moral dilemmas we have selected. It was helpful to the writing process, but also slightly concerning to all three of us, that we gathered the material for these cases so easily. The cases practically fell into our laps everyday as we were writing. And we left many, many examples out—there were so many cases and examples out there that we simply could not include them all or this book would be thousands of pages long. The numerous news articles and examples we mentioned throughout the book, as well as those we left out, were all about ethical problems involving digital technologies. This plethora of material suggests that, as a society, we are seemingly still coming to grips with digital technologies and how to ethically use them. We need to start paying more attention to how our real world lives and our technologies interact and how the short term decisions we collectively make about technology play out in the long run: By understanding digital ethics, we can become less *reactive* to ethical issues and more *proactive* in our approach.

Without strong moral grounding rooted in principles we as a society collectively support, these technologies can quickly get out of hand. And new trends in cutting-edge technology research such as AI, machine learning, AR/VR, and neural networks create fertile breeding grounds for exciting but ethically concerning new ideas and applications to develop. For example, we can think about the technological landscape that might exist a decade or two into the future and consider questions like "What rights and responsibilities do autonomous AI programs have?" and "Is it okay to do terrible things in ultra-realistic simulations if it's all happening virtually, with seemingly no physical consequences?" Questions like these have been explored extensively in science fiction novels and films, but for the first time in our lives, the scenarios they describe are now imminently possible with the technologies we have today and the technologies possible on the near horizon. Certainly we need strong minds focused on the ethical dimensions of these emerging technologies.

As we each bring our own interdisciplinary backgrounds to bear on cases in digital ethics, we come to realize that an interdisciplinary expertise is necessary to become more richly literate in this domain. We need sustained engagement from psychologists, sociologists, legal experts, medical doctors, artists, engineers, scientists, philosophers, humanists, and developers, coders, and technicians. And we need to better understand the working conditions of not only of white collar employees, but also assembly line technicians, construction workers, childcare workers, and the thousands of other professions touched by digital technology, whether these people are affluent or living below the poverty line. In other words, we do not predict a dearth of research opportunities in the area of digital ethics nor a paucity of stakeholders impacted by ethical dilemmas in digital technology. Despite a recent influx of books on this topic, several of which are discussed above, we still see tremendous value and potential in future studies in this area and in additional multidisciplinary techniques for engaging with this topic.

Finally, although we express caution and urge our readers to think deeply about the potential ethical problems that might emerge in digital contexts, we also view the future of digital technology with optimism. As evidenced by the data presented in books such as *Factfulness* (Rosling, Rönnlund, & Rosling, 2018), society is, in general, trending in a positive direction, with improvements in key statistical measures of quality of life occurring throughout the world. Technological advancements in medicine, materials engineering, communication technologies, and education are responsible for much of this progress. However, this trend toward improvement does not mean that we should simply brush the problematic aspects of technology under the rug. Rather, we should recognize that they are there and know that we have the ability and agency not to be morally diminished or dehumanized by technologies we do not always fully understand or control. Addressing issues of moral complexity, like inequality or injustice, is not something that is easily done, but equipping ourselves with an understanding of how digital ethics "works" is the first step in determining how to seek the appropriate resources to help make the world a better place. And we believe this is a morally laudable goal.

Understanding Digital Ethics

By leaning on the concepts, skills, and literacy-based approach we have developed, this book will prove useful not only as an academic resource that advances our current understanding of digital ethics, but also as an applied resource for considering the ethical implications of the digital technologies and cultures as they pertain to users as individuals, as groups in communities and political structures, and as professionals. Such technologies evolve, in the digital, at greater speed and with greater scope than our human ability to understand them fully. This work argues that tools for comprehensive moral reasoning are critical for understanding the consequences these technologies have on us and our surroundings. Developing the relationships among moral sensitivity, ethical reasoning, and moral motivation—the three central components of moral literacy—offers readers a framework for engaging in critical and constructive discourse, reasoning, and action about contemporary issues of digital ethics. We hope you have found our contribution engaging—and that it has helped you develop both digital and moral literacies for your own developing understanding of digital ethics.

References

Christians, C.G. (2019). *Media ethics and global justice in the digital age*. Cambridge, UK: Cambridge University Press.

Couldry, N., Madianou, M., & Pinchevski, A. (Eds.). (2013). *Ethics of media*. Basingstoke, UK: Palgrave Macmillan.

Davisson, A., & Booth, P. (Eds.). (2016). *Controversies in digital ethics*. New York, NY: Bloomsbury Academic.

Dewey, John. (1929). *The quest for certainty*. London: George Allen & Unwin, Ltd. Retrieved from https://archive.org/stream/questforcertaint029410mbp/questforcert aint029410mbp_djvu.txt.

Elliott, D., & Spence, E.H. (2017). *Ethics for a digital era*. Hoboken, NJ: John Wiley and Sons, Ltd.

Ess, C. (2014). *Digital media ethics*. Cambridge, UK: Polity Press.

Leavitt, A. J. (2018). *Digital ethics: Safe and legal behavior online*. New York, NY: Rosen Publishing Group, Inc.

Milton, J. (1674). *Paradise lost, book 8*. Retrieved from https://www.poetryfoundation.o rg/poems/45744/paradise-lost-book-8-1674-version.

Reyman, J., & Sparby, E.M. (Eds.). (2019). *Digital ethics: Rhetoric and responsibility in online aggression*. New York, NY: Routledge.

Rorty, R. (1979). *Philosophy and social hope*. New York: Penguin Books.

Rosling, H., Rönnlund, A.R., & Rosling, O. (2018). *Factfulness: Ten reasons we're wrong about the world—and why things are better than you think*. New York: Flatiron Books.

APPENDIX: DEVELOPING CASES IN DIGITAL ETHICS

This book has a specifically practical goal: We encourage readers to use it as a starting place to think about other and more recently emergent cases that might help them to better understand digital ethics. This appendix has the explicitly pedagogical aim of helping you to conceptualize problems as cases, structuring cases as a means of engaging in argumentation designed to propose and implement solutions to digital ethical problems. We return to one specific case to which we can apply techniques for developing and assessing cases. We encourage readers to apply these techniques to the cases we present throughout the book, and to new cases they might identify and develop.

To explain the process of identifying a "case" in digital ethics, consider this slightly revised version of the smart home devices we offered in our "Voice Devices and Privacy" case from Chapter 3:

> A quick look at an Amazon account with which an Alexa device is associated will show a listing of every command given to the device including voice recordings from misheard commands that were not intended for the device. Examples include television commercials playing in the background in which the "wake word" or a similar spoken sound is noted by the device, and a short (and sometimes not-so-short) recording of the background television noises and the voices of people in the vicinity of the device capturing their conversations as a result of the device "listening" for the wake word. These recordings appear along with every command intentionally given to the machine, from the command to play a movie, television show, or music to a query regarding the time or temperature. One example noted in the news (Weise, 2018) is of a Portland, Oregon family's private conversation being recorded by an Echo speaker and emailing it to a person from the owner's contact list. The friend from

the list contacted the owners to let them know that their private conversation had been sent to him.

Concerns are raised about privacy, spying, and sensitive information being saved, stored, and sent without the knowledge or consent of the owner. Concerns about the use and capabilities of Amazon Echo and other, similar devices such as Google Home and even smart watches have raised concern in Germany, where smartwatches and Internet connected toys are being banned (Diaz, 2017) due to privacy and safety concerns for children who have them and for the privacy of others (such as parents who are eavesdropping on teachers in schools through their children's connected devices) who may be recorded without their consent. In addition, an August 31, 2018 article (Diaz, 2018) reports that there is a new security hole in Echo devices that will allow a hacker to "hijack" the device.

Note that in this case, *writing the case* has been done simply by explaining and presenting some basic information about the issues you believe may be elicited by it. It may depend upon your expertise and interests what captures your attention, but in this case the security, privacy, publication, storage, and safety of one's private information—and the safety of one's person and children—are made prominent by the presentation of at least the following: (1) unintentional recording of conversations and associated retention of recordings, (2) sending private information to an entry in one's contact list, (3) Germany's concern with children's safety and privacy (and the U.S.' different treatment of concerns with Internet-connected devices), and (d) a new hack that does not appear avoidable. In the process of fleshing out the cases (the subject of the next section in this appendix), additional problems may be revealed that the original case presentation did not contain.

An effective way to explain the process of identifying a "case" in digital ethics is to present information from a news or social media source for illustration, as cases from this book demonstrate. There is a wealth of information in such cases which lead to both legal and moral issues concerning their uses. Even after recognizing that a digital ethics case (or any other ethical topic content) may be affected and influenced by the interests or concerns of the speaker or writer who brings issues to your attention, both you and the speaker/writer have an obligation (a moral one) to delve more deeply into the issue to be sure that you are considering the case from multiple angles. To do so allows you the opportunity to identify one issue from more than one point of view and perhaps also to identify other issues that may not have been immediately evident. This is one of the benefits of a pluralistic approach to ethics and the analysis of moral problems.

Actually "writing" a case may not be necessary depending upon the context in which it is being considered or discussed, but it can be useful if your goal is to provide as much information as possible to inform others of relevant facts. It is like the difference between hearing vaguely about how last night's football game turned out and actually reading an account of the game in the sports section of the daily newspaper. The sports reporter notes and writes the relevant details of the game, which players' plays were noteworthy, and so on. But the newspaper article about the game is not a transcript of literally everything that happened during the game. Having as much information as possible would make the newspaper article interminably dull and most of its contents unimportant (for example, would it matter how many hot dogs were sold to the fans during the game?). In the same way, your write-up and presentation of a case in ethics must be carefully considered. If you have too much information, it could make your presentation of a case unwieldy. If you do not have enough information, your case is unclear. For example, consider again the case of Conrad Roy, the young man who committed suicide after receiving a text message from his girlfriend in which she encouraged him to do so. What if the case was simply presented like this: "Conrad Roy killed himself from carbon monoxide poisoning after having a text message conversation with his girlfriend"? No moral issues regarding his girlfriend's words in the text messages would be available to a reader to elicit a moral issue arising from the content of the messages.

Working in teams made up of mixed expertise is also important. One single person may identify a particular and problematic moral issue in digital ethics that overshadows other, and perhaps equally important, issues for which the original presentation of the problem could not account. When all team members or people having a stake in the development, use, sale, and distribution of digital technology are free to look for, to discuss, and to propose solutions to potential and actual problems, it is—at least ideally—the case that some problems can be avoided. Others can be managed to minimize harm or damage, and still others can be solved or lead to the necessity of removing a particular digital technology from use until the problems (depending on their severity) can be addressed and resolved.

Identifying and Mapping Facts

Once a case has been identified and either written out or explained to others, the process of setting out facts and issues and engaging in background research is necessary. It is rarely the case that all the information one can need to make a decision on a moral issue is presented in a written or verbally communicated case. Again, a person interested in the safety of children may be interested in the problem of a (so far) irreparable security hole in Echo devices, but not as much as she or he is interested in keeping the devices out of the hands of children. When one's interests are narrowly focused—and this is certainly not necessarily an undesirable thing—it is often because that person is a specialist in a particular field or aspect of a field. This is again why it is important whenever possible to work in groups or in communities of shared interest to devise means of solving problems as they arise. So, in the case of Echo devices and others like them, a person with interest or specialization in privacy will likely see more completely one aspect of a case while another with interest or specialization in the ethics of hacking will have information and ideas not immediately apparent to others.

When thinking of the issues involved in a digital ethics case, one may profitably utilize a method of setting out problems and arguments in which a three-stage process of addressing a case is used. The method involves "mind maps" or argument structure diagrams to conceptualize a case as (1) factual elements presented in the written or spoken case, (2) questions to ask to gain more information about the case, and (3) arguments offered in favor of a conclusion.

Mind mapping and argument structure diagrams are one effective means, among many possibilities, of identifying facts, formulating questions, and proposing and arguing for solutions to moral problems. We think they are particularly useful in digital ethics both because they offer visual representations of the complex relationships between ideas and arguments but also because many good digital tools exist for doing this kind of mapping—and digital tools for doing digital ethics seems right. Illustrated in Figure A.1 are concept maps used for each purpose in the Amazon Echo/Alexa device case.

1. Every command given to Alexa/Echo devices is recorded in Amazon accounts, including commands that were not actually intended for the devices to "hear."
2. Some recordings included background noises and conversations of people near the devices.
3. An example of unintentional recording by a device was in 2017 when a family's personal conversation was recorded and then sent to a person on the family's contact list.
4. Germany has banned the use of "smart" watches and Internet-connected toys for children due to safety and privacy concerns both for children as well as for others, such as teachers, whose words in class may be recorded or transmitted by the devices without their knowledge and consent.
5. It is possible for hackers to hijack Internet-connected devices like Amazon's Echo. It appears that the security hole cannot be fixed.

It is up to the user to decide whether to identify the facts and associated elements of a case with letters, numbers, symbols, or words and phrases. In the maps above and below, words, phrases, and numbers are used to identify parts of a case.

From the written or verbal presentation of a case, one may conceptualize the facts (from which questions arise and the need for further research is discovered). In the case about Amazon Echo devices, the facts can be separated from each other and thereby more easily organized. So, for example, in the Fact 5 node of the map, one could place with the main fact regarding hacking that there is an article referenced in the case regarding the ease with which the device is hacked. Associated facts would be placed in the Fact 5a (or additional nodes for facts associated with hacking from the information presented). Fact 4 (or any other number) could be used for issues involving the safety and security of children, with

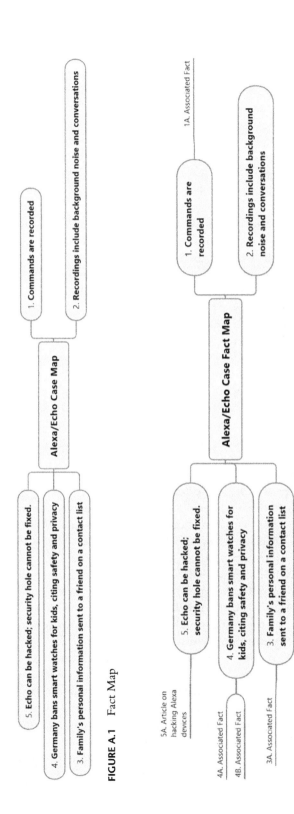

FIGURE A.1 Fact Map

FIGURE A.2 Additional Facts Map of the Alexa/Echo Case

additional and subsidiary facts being placed in Fact 4a and Fact 4b (and other) nodes, and so on for other facts of the case. It is even possible that for a topic/fact, there are no associated facts that you have identified. But when you do, they are easily added, as illustrated in Figure A.2.

The previous list of facts presented in the case can be used to consider questions that you and your associates may wish or need to investigate regarding it. Some possible questions can be put into the list in sub-topics, such as the list below. (It is important to note that you may think of additional questions that are not represented in the example.)

1. **Every command given to Alexa/Echo devices is recorded in Amazon accounts, including commands that were not actually intended for the devices to "hear."**
    a. Are users of the devices properly and clearly informed that devices will record parts or all of conversations if the devices recognize or mistakenly recognize the associated wake word?
    b. Are users also informed that the recorded information appears in their Amazon accounts?
    c. Is it possible for users to turn off the "listening" capacity of their Echo devices, short of turning the devices off completely?
2. **Some recordings included background noises and conversations of people near the devices.**
    a. Is it possible to permanently erase the recordings made by the devices?
    b. Do users of the devices have the right to privacy with respect to the information and sounds recorded by the devices?
        i. Do the providers of the devices (such as Echo) have an obligation to protect the privacy of users?
            1. If they do have such an obligation, how is it done?
            2. Are there limitations to the obligation of providers (such as Amazon) to protect the privacy of users, if Amazon and other providers have any such obligations at all?
    c. What is the obligation of users of the devices to be informed about the capabilities of Internet-connected devices they use?
3. **An example of unintentional recording by a device occurred in 2017 when a family's personal conversation was recorded and then sent to a person on the family's contact list.**
    a. Can the capacity to send information outside the device be disabled?
    b. Should the device be sold with the capacity to inform users when their conversations are being recorded and information is being sent?
4. **Germany has banned the use of "smart" watches and Internet-connected toys for children due to safety and privacy concerns both for children and for others, such as teachers, whose words in class may be recorded or transmitted by the devices without their knowledge and consent.**

    a.   Is Germany's ban on Internet-connected devices for children appro-priate in other government systems, such as the U.S. or Britain, for example?

5. **It is possible for hackers to hijack Internet-connected devices like Amazon's Echo. It appears that the security hole cannot be fixed.**
    a.   Why is it impossible to fix the security hole in the devices?
    b.   If it is not possible to fix the security problem, what is the moral obliga-tion of the device's manufacturer? What is the obligation of the owner of the device?

Each of the facts and associated questions in the case can be represented in a concept map, from which, ultimately, information from the sources of informa-tion you use to answer the questions and the arguments you create to argue for a position, will be placed. (For a more detailed discussion of concept mapping in ethical argumentation from which our discussion here derives, see Stanlick & Strawser, 2015). It is also possible to design maps differently based on their size and complexity. As the information in the map becomes more dense or complex, a different map shape or format may be used, as shown in Figure A.3, which was created with the software program MindManager 2019.

    We have presented mind maps as a method of conceptualizing and presenting cases, questions, and proposed solutions; but we have also noted that there may be cases in which a case is too small or too large for ease in using concept map-ping, or there may be people for whom conceptual maps are not clear or useful. In cases such as these, there are alternate methods of case, question, and argument presentation that may be as profitably used as conceptual maps. Further, concept maps need not be created with specialized software. They can be written by hand or with textboxes in Microsoft Word or another word processing program.

    Another method of case presentation is simply to list the facts and secondary claims in a bulleted or numbered list in whatever word processing software one prefers or to tabulate information by using a table created in a spreadsheet. It is also possible to present cases in some instances simply by verbally communicat-ing with others, by writing information on a whiteboard, sending an email mes-sage, or jotting down a few notes on a piece of paper. The point is that whatever method of conceptually presenting and understanding cases in digital ethics is preferable to you, that you follow the principles of good reasoning. Practicing due diligence in obtaining reliable information from reputable and authoritative sources to argue for your position following the principles of good reasoning will help to create the best argument.

    Note that in the concept maps presented above, the questions about the facts of the case are separated from the larger, original map. Some people consider-ing the case may find it easier to think of the questions and problems of a case in a visual image like a concept map, while others may prefer a list to which they can add content. Others may devise their own method of conceptual-izing a case that is different from either a numbered or bulleted list or from a

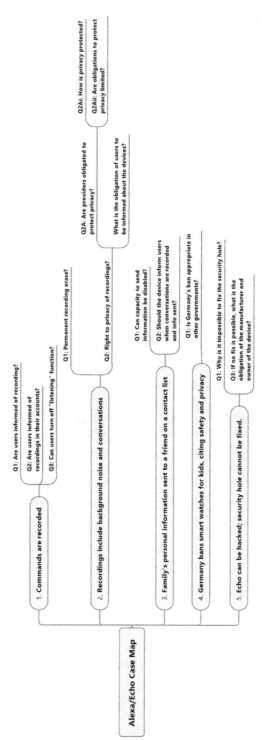

FIGURE A.3 Question Map

concept map. For our purposes, using lists (which can be edited easily to add questions, comments, problems, etc.) developed in word processing software is done by hand just as well as done digitally. Concept maps, on the other hand, while possible to create by hand, are better and more neatly created using digital software tools like Mindjet's MindManager (2019) or freely available software such as MindMeister (2019) or even text boxes with arrows and lines in Microsoft Word. And what better method to use in topics on digital ethics than digital tools?

The point, regardless of method, is to make the issues clear. You should choose an appropriate strategy that works for you to set out cases, facts, questions, and to identify and solve moral problems. This will ensure that you and those with whom you are working have reasonably fully analyzed the case and its issues, and that doing so will facilitate the process of motivating you and your associates to find the means to solve the problem(s) you have identified.

It is always the case that there is more to be learned about an issue than what appears in the presentation of a case. With respect to hacking, for example, it may be relevant to understanding the case and identifying its problems that there are sometimes cases of benevolent (White Hat) hacking, that there are often malevolent (Black Hat) ones, and perhaps further, that there are different means by which successfully hacking into a device may be achieved. With respect to the safety of children, one may need to delve more deeply into the issue of the use of smart watches to find out what sorts of functionality they possess, how the information obtained from smart watches is stored and used by first- and second-parties, and whether owners and users are able to control the distribution of their personal information stored in the devices.

Obtaining more information about the nuances and specific details of a case will arm the investigator with additional facts and perhaps even the identification of problems that other individuals involved in the analysis of a case may not have considered on their own. How much information to seek, and from which sources, are questions beyond the scope of the technique of mapping facts, but it is certainly not beyond the scope of a discussion of the importance of accurate, quality, and authoritative information obtained in arguing for one's own position.

The Importance of Research

In setting out the epistemic facts of the case in a manner that makes the main and secondary ethical issues clear, questions to ask regarding the case should come to the forefront of consideration. In the case of Echo and other Internet-connected devices, it may be that a single investigator or a group of interested people decide that the primary moral question of the case is whether parents ought to buy and use such devices for children. It is not necessary that this question become the primary question of the case—this will depend on the interests and purposes of those investigating the case, and whether there may be some other, more over-arching issue that is of concern to them or that arises while researching a case. An example of another potential primary case question is that of the moral obligations of developers and distributors of the devices to be more diligent in informing the users and the public about vulnerabilities of the devices and the people who use them. Another is the moral obligation of users in protecting their own information and privacy with respect to device use, among others.

With each (and again it is possible to have more than one) primary question of a case comes secondary questions. If you (and your collaborators) have decided that the primary question of the case involves the moral responsibility of developers of devices to inform users of the devices' vulnerabilities, it does not detract from the importance of other associated questions. These may include questions such as those already mentioned regarding the moral responsibilities of parents or the responsibilities of governments or software developers, and what any of these stakeholders ought to do, and so on. No matter how you and your collaborators decide to conceptualize the case, the important issue in rounding out your analysis of a case is to decide what questions you need to ask to be able to investigate as fully as reasonable the issues at hand.

Perhaps a good question to ask is how reasonable it is for users to expect developers to be required to be aware of every "side-effect" or unintentional effect of the use of their devices. Another reasonable question in the Echo case is whether it is possible to close all the "holes" in programming and render the devices completely secure from hacking or from undesired dissemination of personal information. Asking such questions about the case will lead you, if you are careful and diligent, to seek out articles and books on the issues and questions discovered, resulting in the ideal case of the investigators being more fully informed and therefore more fully capable of offering reasonable solutions to the moral problems involved. If, for example, an investigator wrongly believes that it is possible for a software developer to create software that is completely immune to hacking or to undesirable uses, it will be impossible to offer solutions to the problems of Internet-connected devices that are practically possible, short of making the obviously questionable and unjustifiable pronouncement that it is not morally acceptable for anyone to use such devices at all.

Another issue to consider in researching the questions and facts of a case is the division of labor among those involved. Because we present a pragmatic and pluralistic approach to the analysis of moral problems, we hold that it is best when concerned parties investigate the facts, questions, and issues in a case by dividing research labor.

A member of your group who specializes in the development of software might investigate the potential for disabling recordings; another member with knowledge of security might work with questions regarding the security "hole" in the devices; and yet another member with interest in and knowledge of sociopolitical concerns about privacy and business practices may investigate the issues of obligations regarding users' rights, privacy, and the permissibility of banning the use of Internet-connected devices by children. Having various members of a group investigate different elements of the case allows each to focus his or her attention closely on specific issues and for each group member's questions and answers to be combined and synthesized into a coherent argument. We recognize that it is not always possible for groups of people to work together on every problem. When it is not possible, individuals can do well on their own. Nonetheless, variations in experience, knowledge, expertise, interest, and ability can be used to maximum benefit in approaching problems, arguments, and solutions. Pluralism in case design is just as important as in reasoning through case studies.

Arguing for a Position: Reasoning and Thinking Through Cases

As we have noted, a good argument is composed of accurate facts, relevant information, and a sufficient amount of information. Arguments are composed of premises and conclusions. Some conclusions are main or primary points that you and your collaborators wish to put forth, while other conclusions are intermediate ones. Intermediate conclusions serve a dual purpose, being both conclusions from one or more other premises, and being in turn reason(s) for further conclusions, whether intermediate or final (main) conclusion in an argument.

To begin to construct your argument, it is necessary to identify what you believe to be the primary moral question(s) of the case. Suppose that you (and your collaborators) contend that there is one primary question for the case, and that it is this: What is the moral obligation of parents in buying for and permitting their children to use Internet-connected devices such as Amazon Echo devices and smart watches? Even though this may be the main question you have identified at this point, your research and subsequent framing of the case may lead you to a different conclusion than that which flows directly from the original primary question. This is an indication of the dynamic nature of argumentation, and the importance of the questions you ask about the case. They lead to refinement of your thoughts about a case, which in turn may lead you (though it is not necessary) to a different primary question and conclusion from your original inclination about the case.

Suppose, further, that you have answered the factual questions presented earlier in Figure A.3 and are satisfied with the information gathered. Because it is possible and desirable for *you* to investigate more fully the Echo case in this chapter—both with respect to factual and moral questions and concerns—we are presenting only basic information on a potential argument that you may construct to support a position on the question: The obligation of parents in buying and permitting their children to use Internet-connected devices is to be sufficiently informed of the device capacities, especially with respect to the potential for breaches of privacy and security with respect to children, and that, ultimately, it is inappropriate for government to ban such devices for use by children.

You can use a mind- or concept-map technique to create or represent an argument about this case like you did for the facts of the case above. Breaking down the parts of the argument, both visually and verbally representing the inferences made, allows you to identify points in your own argument in which you may need further explanations. Such instances reveal where you may need to use another moral principle, where you require additional facts, and where your argument may commit some error in reasoning. It is possible to find such factors of your argumentation without the use of a concept-based argument map, but it is often easier to note relationships between concepts and statements and to identify needed elements to support your claims with the use of concept mapping.

Remember that your goal in arguing for a position is to provide relevant information, enough information, and statements that are true, clear, and verifiable. If your argument fails to satisfy any of these requirements, it will suffer weakness and lack cogency. Using concept maps of individual inferences or of entire arguments that show individual inferences is a means by which to clarify argumentation and to ensure that the argument you create and present satisfies the relevance, strength, and truth requirements of good reasoning.

The "truth" requirement of a good argument is both the first and the last issue to consider in evaluating and constructing arguments. If your intention is to fool people into believing that something is true, you may not care whether the claims made in your argument are true. If it serves your purposes to have people believe that the entire electoral system in the U.S. has been hacked and made meaningless, for example, then you may be tempted to believe any claim made that appears to justify what you wish to believe. But just as the old adage "wishing doesn't make it so" is learned early in our lives, so too "believing doesn't make it so" should be one of your first lines of defense against supporting and spreading falsehoods. So, in the analysis and construction of arguments, it is incumbent upon the honest investigator to perform background research on claims being made when questioning their veracity is reasonable, and it is also our responsibility as people on a quest to find and disseminate accurate information to exercise caution in making claims that are not obviously or not reasonably to be taken as true without further investigation. It is reasonable to take care and practice diligence in ensuring as a moral requirement that you use only accurate information in an argument you create, and to demand it from others in theirs.

In working through fact and argument mapping, you will notice that there is much more to any argument than simply to say that X is your position without arguing for it. This is true especially since there are people who will disagree with you, and if the issue is as important as it seems to be, careful mapping and identification of elements of a case will assist in identifying facts, arguments, and reasons that will drive analysis of the case.

List of Case Studies in the Book

You can use techniques like these to assess cases we have offered throughout this book—we will list those for you here:

- Text-Assisted Suicide? (Ch 2)
- Voice Devices and Privacy (Ch 3)

- Major Data Breach (Ch 3)
- Active Shooter (Ch 3)
- Handlebar Henry (Ch 4)
- Unreal Elderly (Ch 5)
- My Best Friend is a Robot (Ch 5)
- Bill the Time Traveler (Ch 6)
- Cyberbullying (Ch 7)
- Games and Desensitization (Ch 7)
- "Tay-ken" Over (Ch 8)
- The Dakota Access Pipeline and Social Media Activism (Ch 9)

The difficult ethical work in working through cases studies in digital ethics is two-fold. First, it is determining the balance of harms and benefits (consequences) as well as the values to which we are committed (deontological principles). Second, it is developing a factual understanding of the nature of the digital, its technologies and cultures. But, as we argued throughout this book, this is work essential to understanding digital ethics. Enjoy the process!

References

Diaz, J. (2017, November 12). When is the U.S. going to ban the internet of things for children? *Fast Company*. Retrieved from https://www.fastcompany.com/90151786/when-is-the-u-s-going-to-ban-the-internet-of-things-for-children.

Diaz, J. (2018, August 31). Alexa's alarming new security hole may not have a fix. *Tom's Guide*. Retrieved from https://www.tomsguide.com/us/alexa-skill-squatting-hackers,news-27946.html.

Mindjet.com. (2019). *MindManager*. Retrieved from https://www.mindjet.com/mindmanager/.

Mindmeister. (2019). *Mindmeister*. Retrieved from https://www.mindmeister.com/.

Stanlick, N., & Strawser, M. (2015). *Asking good questions: Case studies in ethics and critical thinking*. Indianapolis: Hackett.

Weise, E. (2018, May 24). Alexa creepily recorded a family's private conversations, sent them to business associate. *USA Today*. Retrieved from https://www.usatoday.com/story/tech/talkingtech/2018/05/24/amazon-alexa-creepily-recorded-sent-out-familys-conversations/642852002/.

INDEX

For Product Safety Concerns and Information please contact our EU
representative GPSR@taylorandfrancis.com
Taylor & Francis Verlag GmbH, Kaufingerstraße 24, 80331 München, Germany

www.ingramcontent.com/pod-product-compliance
Lightning Source LLC
Chambersburg PA
CBHW071425050326
40689CB00010B/1988